FIRE DEPARTMENT
INCIDENT SAFETY
OFFICER

Online Services

Delmar Online
For the latest information on Delmar Publishers new series of Fire, Rescue and Emergency Response products, point your browser to:
http://www.firesci.com

Online Services

Delmar Online
To access a wide variety of Delmar products and services on the World Wide Web, point your browser to:
> **http://www.delmar.com**
> or email: info@delmar.com

A service of I(T)P®

FIRE DEPARTMENT INCIDENT SAFETY OFFICER

David W. Dodson

Delmar Publishers

an International Thomson Publishing company

Albany • Bonn • Boston • Cincinnati • Detroit • London • Madrid
Melbourne • Mexico City • New York • Pacific Grove • Paris • San Francisco
Singapore • Tokyo • Toronto • Washington

NOTICE TO THE READER

Cover/insert photos courtesy of Evan Lauber

Emergency Management Online College Courses, Disaster Planning: http://beat1.spjc.cc.fl.us/em/em.html

Delmar Staff
Publisher: Alar Elken
Acquisitions Editor: Mark Huth
Developmental Editor: Jeanne Mesick
Production Coordinator: Toni Bolognino

Art and Design Coordinator: Michele Canfield
Editorial Assistant: Dawn Daugherty
Marketing Manager: Mona Caron

COPYRIGHT © 1999
By Delmar Publishers
an International Thomson Publishing Company, Inc.

The ITP logo is a trademark under license.

Printed in the United States of America

For more information, contact:

Delmar Publishers
3 Columbia Circle, Box 15015
Albany, New York 12212-5015

International Thomson Publishing Europe
Berkshire House
168-173 High Holborn
London, WC1V7AA
United Kingdom

Nelson ITP, Australia
102 Dodds Street
South Melbourne,
Victoria, 3205 Australia

Nelson Canada
1120 Birchmont Road
Scarborough, Ontario
M1K 5G4, Canada

International Thomson Publishing France
Tour Maine-Montparnasse
33 Avenue du Maine
75755 Paris Cedex 15, France

International Thomson Editores
Seneca 53
Colonia Polanco
11560 Mexico D. F. Mexico

International Thomson Publishing GmbH
Königswinterer Strasße 418
53227 Bonn
Germany

International Thomson Publishing Asia
60 Albert Street
#15-01 Albert Complex
Singapore 189969

International Thomson Publishing Japan
Hirakawa-cho Kyowa Building, 3F
2-2-1 Hirakawa-cho, Chiyoda-ku,
Tokyo 102, Japan

ITE Spain/ Paraninfo
Calle Magallanes, 25
28015-Madrid, Espana

2 3 4 5 6 7 8 9 10 XXX 04 03 02 01 00 99 98

Library of Congress Cataloging-in-Publication Data

Dodson, David W.
   Fire department incident safety officer / by David W. Dodson.
      p.   cm.
   Includes index.
   ISBN 0-7668-0362-7 (paper)
   1. Fire extinction—Safety measures.   2. Fire fighters.
   3. Safety engineers.   I. Title.
   TH9182.D63 1998
   628.9′25′0298—dc21   98-3046
                                                                CIP

This book is about pride, professionalism, knowledge, and attention to detail. These attributes will ultimately make a difference in the fire service. One man, more than anyone, has helped me understand this. Therefore, I lovingly dedicate this book to that man, my dad, H. C. "Chuck" Dodson.

Contents

Preface

This book has one main purpose—to reduce firefighter injuries and fatalities. Now this may sound noble, but I actually have a take on firefighter injury and death. I believe that having a trained and experienced Incident Safety Officer is the most tangible, cost effective, and prudent way to reduce firefighter injury and death and is one of the fastest solutions we, as a fire service can implement. Approximately half of our fatalities and injuries occur on the incident scene. So what can we do right now, with virtually no cost, to reduce these injuries? Simple—appoint a trained safety officer to help the Incident Commander take care of our comrades.

Currently, a few outstanding training courses exist to assist the potential Incident Safety Officer in refining his or her skills. Additionally, several fire service texts address the Incident Safety Officer role and responsibilities as part of other curriculums (i.e., Incident Command, Risk Management). This book goes into more detail and specifically addresses the conditions, signs, warnings, and hazards that the Incident Safety Officer must be alert to in order to prevent injury. Further, the book presents a system to help the Incident Safety Officer keep track of everything that goes on during an incident and prioritize what needs attention.

While writing this text, I kept a focus and vision that the contents be extremely realistic and practical, something a company officer, sector officer, incident commander, or, for that matter, any firefighter could read and use in their respective position. After all, it takes a team effort to handle emergencies and accomplish tasks safely. So, while the focus here is on what the Incident Safety Officer needs to do, the observations, evaluations, and actions suggested can apply to any hands-on, working member of a response team.

Great consideration was given to the vernacular and acronyms used in this book. Just the title of Incident Safety Officer (ISO) creates acronym conflict (for example, with Insurance Services Office). The National Fire Protection Association (NFPA) standards and the National Fire Academy have chosen *ISO* and *HSO* to represent Incident Safety Officer and the Health and Safety Officer respectively. The National Interagency Incident Management System (NIIMS) chose Safety Officer Type 1 and Type 2 (SOF1, SOF2). I considered using Duty Safety Officer (DSO), Incident Scene Safety Officer (ISSO), and Emergency Safety Officer (ESO), as well as the ISO or HSO and SOF1 and 2 acronyms. Of all these choices, it was felt that the title Fire Department Incident Safety Officer (FDISO) most closely captured the nature of the position and avoided conflict with other acronyms. This acronym also comes closest to the language used in National Fire Academy programs and NFPA standards.

Regarding command systems, I tried to use the common language of the National Fire Academy Incident Management System, the consortium language, as well as the Fire Command

(Brunacini) system. It is hoped that the reader does not get hung up on "proper" command language—the information here can fit into *any* system.

ORGANIZATIONAL FEATURES

This book is divided into two sections. Section 1 prepares the incident safety officer for his or her duties. In order to make evaluations and judgments in the field during the urgency of an incident, the incident safety officer needs to have a solid foundation of knowledge in fire behavior, building construction, hazardous energy, human performance, and risk management. Likewise, a system of getting an effective incident safety officer on scene is necessary. Section 1 introduces these foundations. Section 2 presents the *Incident Safety Officer Action Model*, which gives the incident safety officer an easy-to-remember thinking model to help sort, evaluate, and act on the many conditions and operations found at emergency incidents.

Throughout the text, many figures and pictures are included, and key points have been highlighted to help make the book readable and to help you visualize applications during emergencies. Also, we included some teaching aides such as objectives, review questions, and summaries. To further assist instructors, an instructor's outline and resource manual is available from the publisher.

SCOPE AND VISION OF THE BOOK

In addition to providing a practical, tested approach to accomplish incident safety officer duties, this book addresses many of the ISO duties outlined in NFPA Standard 1521, Fire Department Safety Officer (1997 edition). Currently, there are efforts to create a national certification for the Incident Safety Officer. Also, a few state and local systems have created Safety Officer Certifications. This text is intended to help candidates study for these certifications.

SUGGESTIONS ENCOURAGED

This book is reflective of many years of teaching incident safety seminars across the United States and Canada; many people have influenced its content. Likewise, a team of technical reviewers have provided valuable comments and suggestions that have produced a better text for you. As you read this book, I encourage you to envision its application, practice the action model, and provide suggestions and comments so that the next edition will be an even better tool to prevent firefighter injuries and deaths.

Acknowledgments

This project represents more than the effort of this lone author. In fact, it would take many pages to thank all the people who influenced this project. The material here represents more than 20 years of fire service adventures gleaned from teaching firefighters in virtually every corner of the United States, from many hallway gatherings at the National Fire Academy, and from those kick-butt incidents with my fellow firefighters. These adventures have taught me more than any book could ever present.

As with any venture, a starting line must be crossed. For me it started as a wide-eyed kid who loved going to dad's fire station in Arvada, Colorado. I became a firefighter for the U.S. Air Force and was shipped to Elmendorf Air Force Base outside of Anchorage, Alaska. Here, my first supervisor, Sgt. Tony Schultz pointed me in the direction of contribution, pride, and progress. I can never thank Tony enough for his insight, mentoring, and positive outlook. He truly helped me step out on the right foot.

Along a 20-year path, I have been fortunate enough to meet some truly outstanding individuals who encouraged, contributed, and influenced my fire service passion. Chief Duncan Wilke (retired) of the Parker (Colorado) Fire District opened many doors and had the faith that this young trooper could become a training officer. Chief Wilke introduced me to the International Society of Fire Service Instructors (ISFSI) where I became involved with some fire service movers and shakers. Specifically, I would like to thank Daniel B. C. "Boone" Gardiner (Fairfield, Conn.), Larry Davis (IRI, Hartford, Conn.), and Roger McGary (Montgomery County, MD) for all the advice, support, and encouragement. These gentlemen also had the insight to create and support an upstart organization, The Fire Department Safety Officer's Association (FDSOA). In the FDSOA I found a home and met some truly amazing safety officers. Namely, and with thanks, I would like to recognize T. Randy Hess (VFIS), Ted Bateman (Philadelphia, Pa.), Dave Abner (Louisville, Ky.), the past and present FDSOA board of directors, and Mary McCormack (FDSOA Executive Director) who, along with the aforementioned, took a risk with a young and naive instructor.

The FDSOA path introduced me to a couple of energetic fire officers who became great teaching partners and significant contributors to this endeavor. Battalion Chief Terry Vavra of the Lisle-Woodridge Fire District (Ill.) jumped on my mini-band wagon and helped teach this message to over 1,200 firefighters around the United States, and contributed to some of the core ideas herein. Chief Health and Safety Officer David Ross of the Toronto Fire Department also became an ally in this project. Special thanks also to Chuck Soros, Rich Maddox, Phil Chovan, and Frank Bachman.

NFPA standards are the direct result of many energetic and caring professionals. My involvement with the NFPA 1500 group exposed me to four outstanding individuals whom I wish to thank. The safety officer's Safety Officer is Murrey Loflin (Virginia Beach, Va.) who helped me focus on many safety officer issues. Stephen Foley, an NFPA staff liaison, and Bob Neamy (Los Angles City Fire) showed me the NFPA way and coached my contributions. And also in this group, I cannot proceed without acknowledging the contagious love and concern for firefighter safety shown by Angelo Catalano, Jonathan Kipp, and Don Burns.

Cindy Odel is a contract manager for U S WEST who helped me understand the legal and business portion of copyright, contracts, and publishing. Thanks Cindy. Pat Skahill, formerly with the City of Loveland (Colo.), is a risk manager who taught this rookie about the intricacies of worker's compensation insurance and the art of applied risk management. Lieutenant Dave Legits of Loveland Fire and Rescue introduced this city boy to the wildland fire arena and reviewed content for the wildland fire sections of this book. Many of the photos in the book were contributed by the talented Rick Davis, a Lieutenant with Loveland Fire And Rescue. And thanks to the Loveland Firefighters who appear in many of the photos in this text.

The gang at Delmar Publishing has been especially supportive and helpful and, obviously, made this endeavor a reality. Mark Huth and Dawn Daugherty are project masters who steered this new author down the right path. This author and the Delmar gang would like to also show gratitude to the following review group:

Richard Powell
Saginaw Township Fire Department
Saginaw, MI

Mike Flavin
St. Louis Fire Academy
St. Louis, MO

Dennis Childress
Orange County Fire Authority
Ranch Santiago College
Orange, CA

Robert Klinoff
Kern County Fire Department
Bakersfield, CA

Thomas Wutz
New York State Fire Academy
Albany, NY

Last, and most importantly, I would like to thank a few special people who have always believed in me and stepped to the plate when I needed a word of encouragement or a kick in the pants. Battalion Chief Bob Baker (Parker Fire District, Colo.) is my fire service brother. Rich Bristol is my twin brother of a different mother. Marilyn Crafts, formerly of FDSOA, and her husband Peter pushed me past the obvious and helped me see the big picture. And finally, a special thanks to my talented partner and wife LaRae, who, more than anyone, made this book possible with her proofreading, commitment, unwavering love, and dedication to me and our two incredible children Kelsie and Dan.

About the Author

David W. Dodson is a Lieutenant and Duty Safety Officer for Loveland Fire and Rescue in Colorado. Dave has served more than eight years as a Training and Safety Officer for Loveland and the Parker, Colorado, Fire District. He represents the Fire Department Safety Officers Association on the NFPA Fire Service Occupational Safety Committee and chaired the task group that produced the 1997 edition of NFPA 1521, Fire Department Safety Officer. Dave has an associate's degree in Fire Science and an adult education teaching credential. Dave enjoys his family and the Colorado outdoor life when he is not working or traveling the country teaching Incident Safety Officers.

Photo by Don Mackey,
New Image Photography.

PREPARING THE INCIDENT SAFETY OFFICER

Section 1 establishes a foundation of history, knowledge, and process to assist the Fire Department Safety Officer achieve his or her goal in preventing firefighter injury and death. Chapter 1 deals primarily with the history of the safety officer then discusses current safety officer trends and the need to appoint incident safety officers more frequently. Chapter 2 lays the foundation of general safety concepts including risk management. Chapter 3 helps a fire department define or refine its incident safety officer duty system to ensure that a trained fire department incident safety officer (FDISO) is available for response to incidents. Specific technical information such as building construction, fire chemistry, and human performance are presented in Chapter 4. Finally, Chapter 5 discusses the common traps that can render an FDISO ineffective and presents some positive triggers to help the FDISO make a difference.

Chapter

1

Introduction to the Safety Officer Role

Learning Objectives

Upon completion of this chapter, you should be able to:

- Describe the divergence of the safety officer role in fire departments.
- Discuss the history of the fire department safety officer.
- List the NFPA standards that affect and pertain to the incident safety officer.
- Explain the need for an incident safety officer in empirical and image terms.

The fire has been burning for 30 minutes now in a three-story brick build-ing built in the late 1930s. As you complete a 360° look at the place you note the firefighters are just starting to make some headway on the third floor fire. As Safety Officer, you report to the Incident Commander that the building looks OK and, other than a precariously placed ground ladder, you are con-tent, from a safety point of view, with the effort. The incident commander asks you, "What's our collapse potential?" You answer that you thought the ordinary construction is holding up well and that we probably have time to fight fire before it gets weak. Suddenly the radio crackles to life, "MAYDAY, MAYDAY! Command Engine 2, we have a . . . uh, a collapse. Ugh . . . Fire-fighters trapped, send help. MAYDAY!" You both look up and see the fire venting out what appears to be a new hole in the roof. The Incident Com-mander begins making emergency assignments. You look at your SAFETY OFFICER vest and then the Incident Commander. One expression tells the rest of the story.

safety officer
generic title that can mean either an incident safety officer or a health and safety program officer or manager

health and safety officer (HSO)
title given to the person or persons who manage and administer the occupational safety and health program for a fire department

fire department incident safety officer (FDISO)
title given to the officer assigned by the incident commander to handle safety responsibilities and duties

The title "**safety officer**" is used daily in fire departments around the country. Often, this title is used to address the individual in charge of a department's entire safety and health program. Other times, the title is applied to an arriving engine or truck company officer who reports to the Incident Commander (IC) and is delegated the safety officer task (Figure 1-1). In some departments, the safety offi-cer is actually an OSHA compliance officer. In still other departments, the safety officer is another title for the training officer. For this book, we have chosen to think of the safety officer as the individual who leads the entire fire department safety and health program or effort. Further, the safety officer is referred to as the **Health and Safety Officer (HSO).** This book, however, focuses on the one, two, or entire group of firefighters and fire officers who may be delegated the safety task at an incident scene—**the Fire Department Incident Safety Officer (FDISO).**

The reason for splitting titles here is actually a recognition of reality in the fire service. Many fire departments have had, or still have, one HSO who also sup-ports the Incident Commander on incidents. However, a department soon realizes that one individual cannot possibly make every significant incident where an FDISO would be desirable. Incident commanders who like having a FDISO pres-ent found that when the department HSO was not available, they simply chose a person they trusted to fill the void. Thus, a division of the safety officer role began. For clarity in reading this book, take a moment now to refer to Figure 1-2 for some of the roles and responsibilities of both the HSO and the FDISO. As you can see in the figure, the HSO is responsible for health and safety administration whereas the FDISO focuses on scene-specific operations. It is also obvious that some overlap occurs—by design. In overlapping responsibilities, a link is estab-lished to provide consistency and communication in the different roles. Before we delve too deeply into FDISO development, let us quickly review the development of today's FDISO.

Figure 1-1 *An effective Incident Safety Officer can reduce the chance of firefighter injury or death.*

Health and Safety Officer Functions

- Risk management planning
- Procedure review
- OSHA compliance
- Safety and health education
- Data analysis
- Facility and equipment inspection
- Infection control
- Wellness programming
- Accident investigation
- Postincident analysis
- Safety committee participation

Incident Safety Officer Functions

- Risk evaluation
- Resource evaluation
- Hazard identification and communication
- Action plan review
- Safety briefings
- Collapse zoning
- Accident investigation
- Postincident analysis
- Safety committee participation

Figure 1-2 *HSO and FDISO functions.*

HISTORY

Safety officers, in one form or another, have been present in the American work-force for a long time. Some of the first safety officers came out of the fire service. One example, from the late 1800s and early 1900s, were "wall watchers" who stood at corners of buildings and watched the wall for signs of bowing or sagging during a working fire (Figure 1-3). This trend followed the catastrophic collapse of New York's Jenning Building on April 25, 1854.[1] In this tragedy, twenty fire-fighters were buried following a partial, then significant collapse of the building at 231 Broadway. In Colorado Springs in 1898, a decision was made by on-scene officers to withdraw firefighters from a railroad car fire containing black powder. Thirty minutes later, the car blew, causing a wind-fed fire that destroyed many buildings including the famous Antlers Hotel. These are just a few examples of the early safety officer role. In some respects, the fire service was viewed as pro-gressive in risk management.

Figure 1-3 *A late 1800s Fire Officer shouts collapse warnings—the First Safety Officer.*

As America became more industrialized, the need for a safety officer increased—for both fire departments and general manufacturing. Signs of this need were felt in World War I when soldiers became mechanized. It was not until World War II, however, that the formal safety officer was utilized. It seems that the first "industrialized" war brought significant injury and death in *support* operations as well as combat. The military started to look at why people were getting hurt outside of combat, and they began appointing safety officers for an immediate impact while they developed safer processes.

Even as World War II continued, factories and other manufacturing industries began looking at the safety of their workers. Some of this inspection came at the request of the insurance industry, while other safety issues came at the request of organized labor. Before long, safety inspections, posters, briefings, and so forth were commonplace in the manufacturing environment. In 1970, Congress passed the William Stieger Act, which included the Occupational Safety and Health Act (OSHA). President Richard Nixon signed the act into law that December. The law basically gave equal rights and responsibility to employers and employees with respect to safe working conditions. Today you can find even small business with a dedicated safety manager or OSHA compliance officer. In large corporations, an industrial hygienist administers risk management and employee safety programs. Safety in corporate America is so important that many careers have been spawned in the safety arena. Colleges, universities, and even vocational schools offer degree and certificate programs in safety and safety-related programs. One prominent fire chief, Daniel B. C. Gardiner of the Fairfield, Connecticut, fire department, has been quoted as saying that the next major fire service discipline will be that of community and fire department risk management.

FIRE DEPARTMENT SAFETY OFFICER TRENDS

Over the past decade a story has become familiar among training and safety officers. The story goes a lot like this:

> The fire chief went to a national conference and heard some guy talk about the need for a department safety officer. NFPA 1500 says you got to have one. The best person for the job is your training officer. Next thing I know, I'm it.

NFPA

NFPA 1501 has since been changed to NFPA 1521 in an effort to standardize NFPA numbering.

Sounds like a typical fire service story. Unfortunately, not much fire service training material existed to tell the newly appointed safety officer what to do. Most newly appointed safety officers got a personal copy of the 1987 NFPA 1500 *Standard on Fire Department Occupational Safety and Health Program*. A little research found that NFPA 1500 was nothing more than a fire service twist on commonplace practices in the industrial world. Granted, some issues in NFPA 1500 were controversial (staffing, equipment design, etc.) but the basic premise to develop and administer an active health and safety program was its guiding purpose.

As a companion document to NFPA 1500, NFPA 1501 *Standard for Fire Department Safety Officer,* was created by the NFPA Fire Department Occupational Health and Safety Technical Committee. This standard addressed the authority, qualifications, and responsibilities of the safety officer. The original standard addressed primarily the HSO role. NFPA 1501 has since been changed to NFPA 1521 in an effort to standardize NFPA numbering. Both 1500 and 1521 are updated on a regular revision cycle under the guide of NFPA's Fire Department Occupational Safety Technical Committee.

At this point, it is important to note that some safety officer trends were well under way in the fire service. In the 1970s, the FIRESCOPE program was developed and used for multiagency incidents on the West Coast. FIRESCOPE is an acronym for FIre REsources of Southern California Organized for Potential Emergencies. The FIRESCOPE ICS model was adopted into many National Fire Academy courses—leading to widespread adoption of ICS. In the late 1970s, Chief Alan Brunacini of the Phoenix (Arizona) Fire Department began teaching a Fireground Command seminar across the country. In this seminar, it was recommended that a safety officer or "safety sector" be established to provide a higher level of expertise and undivided attention to fireground safety. This sector was designed to report directly to the Fire Ground Commander (FGC) as well as advise and consult other sector officers (Figure 1-4). In 1983, the International Fire Service Training Association (IFSTA) printed a manual titled *Incident Command System.* In this manual, a Safety Officer position was integral to the Command Staff and a checklist and organizational chart were included.[2]

In other examples, large cities such as New York City were creating safety divisions and shift-assigned safety officers to provide injury investigation and incident safety. The **National Interagency Incident Management System (NIIMS),** used by the National Wildfire Coordinating Group, recognizes the safety officer as directly reporting to the incident commander. NIIMS is a direct descendant of the FIRESCOPE program. In NIIMS, the safety officer is classified as either a Type 1 (SOF1) or Type 2 (SOF2). A SOF1 is qualified to deploy nationwide as part of an Incident Overhead (management) Team. A SOF2 is usually a state or local qualified safety officer for wildland and interface fires. Interesting to note here, both the SOF1 and SOF2 must meet the same criteria for qualification.[3]

THE NEED FOR A FIRE DEPARTMENT INCIDENT SAFETY OFFICER

The role of a fire department safety officer is based on a simple premise: We (the fire service) have not done a good job taking care of our own people. Phoenix Fire Chief Alan Brunacini said this best when he said: "For 200 years we have been providing a service at the expense of those providing the service." Proof of his statement is found in the examination of empirical and image factors regarding firefighter injury. With this data, we can show that it is reasonable to utilize an effective FDISO at all working incidents.

■ **Note**
It was recommended that a safety officer or "safety sector" be established to provide a higher level of expertise and undivided attention to fireground safety.

National Interagency Incident Management System (NIIMS)
federally recognized system for managing multijurisdictional emergency incidents

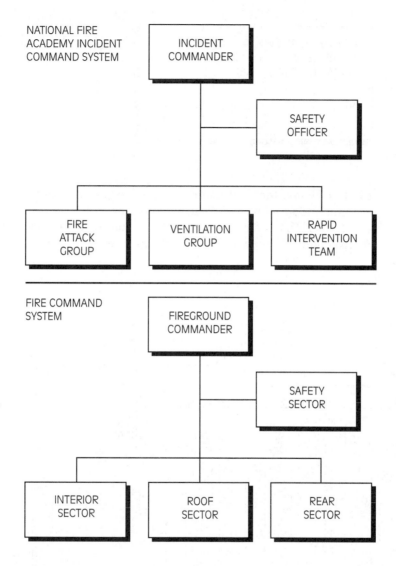

NATIONAL FIRE
ACADEMY INCIDENT
COMMAND SYSTEM

INCIDENT
COMMANDER

SAFETY
OFFICER

FIRE
ATTACK
GROUP

VENTILATION
GROUP

RAPID
INTERVENTION
TEAM

FIRE COMMAND
SYSTEM

FIREGROUND
COMMANDER

SAFETY
SECTOR

INTERIOR
SECTOR

ROOF
SECTOR

REAR
SECTOR

Figure 1-4 *The Incident Safety Officer reports to the person in charge, regardless of the type of incident management system used.*

Empirical Study

Death and Injury Statistics In general terms, one hundred U.S. firefighters are killed every year in the line of duty. This figure has been fluctuating during the last two decades with a high of 171 in 1978 and low of 75 in 1992. However, on average, 118 firefighters have died each year since 1977, with an average of 101 deaths per year since the introduction of NFPA 1500 and 1501[4] (see Figure 1-5). This decline is a positive trend, probably due to multiple factors. Leading the list of factors is the attention brought to firefighter safety through NFPA 1500 and the use of

"FOR 200 YEARS WE'VE BEEN PROVIDING
A SERVICE AT THE EXPENSE OF THOSE
PROVIDING THE SERVICE."

–ALAN BRUNACINI
PHOENIX FIRE CHIEF

safety officers, incident command systems, and other programs and equipment recommended by NFPA 1500. The decline in firefighter deaths may also be the result of OSHA enforcement, liability issues resulting from court action, and risk management programs brought to the fire service by the insurance industry.

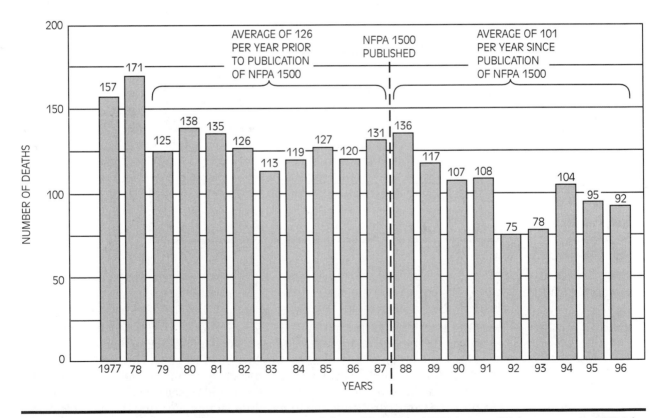

Figure 1-5   *Firefighter duty deaths have decreased since the publication of NFPA 1500. (Figures courtesy NFPA.)*

Approximately half of all duty deaths occur at the incident scene; the remainder of the deaths fall into categories such as responding to and returning from incidents, nonfire incidents, and training. You may argue that this figure is acceptable considering the risks that a community expects firefighters to take. Many fire officers believe that no death is acceptable, especially if it can be prevented by coached aggressiveness. If a single firefighter death can be prevented by the appointment of an incident safety officer, then the effort is worth it.

Firefighter injuries (as opposed to deaths) have declined in recent years. This decline, however, corresponds with a decline in fires. Perhaps a better measurement is the number of injuries per one hundred fires fought. This figure shows no consistent trend (up or down) in recent history.[5] Still, at 45,725 fireground injuries in 1996 (that is 125 per day!), it stands to reason that we, as the fire service, have ground to gain in injury prevention.

Combine firefighter death and injury numbers with a national trend of less fires, and one can hypothesize that the effort to reduce firefighter injuries may not be as effective as it could be. It is easy to see that the appointment of an FDISO, more often than not, seems prudent. Granted, other programs such as firefighter wellness and incident management systems will help reduce injury and death potential over time. The use of an FDISO, can, however, start to reduce these threat potentials today.

An interesting point can be made regarding the need for an incident safety officer. How many firefighter injuries and deaths have been prevented by the action of an FDISO? One Missouri fire chief presented this question when arguing against the cost of formalizing an FDISO program for his department. Interestingly enough, no one present at the discussion could answer the question with hard data, although many felt that they did, in their own experience, change a situation that "could have" led to an injury. In one case, an FDISO at a Colorado strip mall fire called for evacuation of a building and the Incident Commander concurred. The firefighters present withdrew and, after questioning the pullout, witnessed the roof slowly collapse into the building. This incident came just a few short years after a firefighter died following a similar roof collapse in a neighboring department. The point is simple: We do not keep good data on what could have been, and, at times, our memories are too short.

Worker's Compensation When studying the empirical effects of firefighter injury and death, it would be negligent to skip the effects of work-related injuries on worker's compensation. Worker's compensation is statutory for each state; each has its own intricacies. Rates, however, are set by the **National Council on Compensation Insurance (NCCI).** Each state may also adjust rates for firefighters based on experience within that state. This adjustment is called an *experience modifier*. To determine a worker's compensation rate for a given department within a state, a formula of NCCI rate × payroll × experience modifier is used. The experience modifier is based on a 3-year loss experience.

■ **Note**

Worker's compensation programs are not free; they are costly and that cost is quite dynamic based on history—the number of claims and the cost of those claims.

A few points can be made here. First, and most importantly, worker's compensation programs are not free; they are costly and that cost is quite dynamic based on history—the number of claims and the cost of those claims. Second, a fire department cannot always shop around for a good rate like some of us do for automobile insurance. If a firefighter is injured on the job, the ramifications may be felt for many years. Obviously, the more serious the injury, the longer the impact is felt. Further, this loss impacts all employers with employees in the firefighter class. It is easy to imagine the impact such injuries can have on the long-term financial status of a fire department.

Image Study

The image study of firefighter injury and death deals with less tangible results as opposed to data. A firefighter injury requiring hospital care or extended time off creates stress in the workplace. On the obvious side, smaller career departments have to struggle with finding a replacement, while small volunteer departments have to make due without an individual. Large departments have to shuffle people and provide a rover or other fill-in assignment. Less obvious perhaps is the firefighter work slowdown following the accident. Most fire officers have seen this reaction: Concern, introspection, and even trepidation fill the firehouse following a significant injury. The more serious the injury, the more pronounced these displays may be. If an investigation follows a serious injury, the workplace effects of the injury can multiply. Some of these effects include finger pointing and taking sides.

■ **Note**

Concern, introspection, and even trepidation fill the firehouse following a significant injury.

Following a firefighter duty death, labor and management concerns might be expressed in private investigations and attempts to minimize liability. In some cases, career and volunteer officers have been demoted, suspended, and terminated. These events are often covered in the local media, causing additional department, community, and individual stress.

Without a doubt, the most damaging effects of a firefighter injury comes at the expense of the involved families. Stress permeates the family lives of those firefighters who have experienced an injury or have been involved in the incident. Most experienced firefighters have had to justify their involvement with the fire department to a loved one. Usually the firefighter finds him- or herself in this discussion following a news report of a firefighter killed or seriously injured. Many firefighters can recall a peer who has resigned, been divorced, or has chosen to enter an assistance program because of a comrade's injury or death.

The Bottom Line

The message here is simple. The fire service must continue to improve firefighter safety. It is obvious that we are not where we could be. It is with that belief that this book is written. An Incident Safety Officer can make a difference—*RIGHT NOW.*

Figure 1-6 *The creation of an effective Incident Safety Officer program is the Incident Commander's key to incident safety.*

The goal of this book is to give you a systematic and meaningful approach to the creation and implementation of an effective incident safety officer program (Figure 1-6). Further, it is this book's intent to give any firefighter the information necessary to be an effective Incident Safety Officer, or for that matter, a more safety conscious Incident Commander, Company Officer, or Firefighter. Information contained herein can help you tonight, next shift, or in making sweeping changes in your department. The material is that powerful. The remainder of this book is dedicated to the spirit of making a difference.

■ **Note**
Information contained herein can help you tonight, next shift, or in making sweeping changes in your department. The material is that powerful.

Summary

Throughout industrialized history, the title safety officer has been focused on preventing injury and the loss of life through solid risk management and hazard reduction. In the fire service, the term *safety officer* has been generically applied to persons filling administrative as well as incident response duties. Recently, a divergence has taken place and two specialties have emerged: the Health and Safety Officer, which is primarily an administrative or managerial role, and the Incident Safety Officer, which is an incident command staff position.

A strong need exists for incident safety officers because injury and death statistics for on-scene injuries and deaths have not dropped dramatically even though fire departments respond to fewer fires. Further, increasing costs associated with firefighter injuries mandate that more prevention is necessary. The appointment of an incident safety officer on working incidents is a positive step that can help prevent injuries.

Review Questions

1. Explain the divergence of the safety officer role in fire departments.

2. In general terms, explain the history of the modern safety officer in the industrial world as well as the fire service.

3. List and discuss the NFPA standards related to the incident safety officer.

4. Discuss current firefighter injury and death trends and the need for incident safety officer response.

Notes

1. P. R. Lyons, *Fire in America* (Quincy, MA: NFPA Publications, 1976), p. 32.

2. Fire Protection Publications, *Incident Command System,* (Stillwater, OK: Oklahoma State University, 1983), pp. 19–20, 61.

3. National Interagency Incident Management System, *Task Book NFES #2303,* Boise, ID: National Wildfire Coordinating Group (August, 1993).

4. A. Washburn, P. LeBlanc, and R. Fahey, "1996 U.S. Firefighter Fatalities," *NFPA Journal* 91, no. 4, (July/August 1997), p. 48.

5. M. J. Karter and P. R. LeBlanc, "1996 U.S. Firefighter Injuries," *NFPA Journal* 91 no. 6, (November 1997).

Chapter

2

Safety Concepts

Learning Objectives

Upon completion of this chapter, you should be able to:

- List the three elements that affect safety in the work environment.
- Discuss the difference between formal and informal procedures.
- List the qualities of a well-written procedure or guideline.
- Discuss the external influences on safety equipment design and purchase.
- List and discuss the three factors that contribute to a person's ability to act safely.
- Define risk management.
- Identify and explain the five parts of classic risk management.

The lights are down low, the instructor's voice is borderline monotone, and the slide is out of focus. Lunch has settled nicely. You look around at your fellow fire officers who are half asleep. Your mind wanders to the last fire, a two-room and contents fire that went pretty well. You actually got in on the fire attack. Your partner had a question about a building construction feature that you had no answer for. Silently to yourself you say, "Dang, I was going to look that up." Suddenly, the instructor's voice bores into your thoughts saying, "These can kill firefighters if you're not careful!" You rapidly sit upright and try to catch on to the message that was just delivered. "What did he say?"

THEORY VERSUS REALITY

Let us face it, theory can be boring! Most fire officers would just as soon skip the theory or book work necessary to become an incident safety officer, but they crave the practical, challenging, and critical aspects of the assignment. In order to become a fire department incident safety officer (FDISO) who *can make a difference,* one must build a foundation of understanding (theory). Most fire officers agree that an effective FDISO possesses a healthy dose of common sense, but it should also be agreed that an effective FDISO is well grounded in recognized safety concepts (theory), which gives that FDISO "uncommon sense." Uncommon sense can also be defined as that attribute that allows the FDISO to ask "what is the worst that can happen here and what is the probability of that happening?" In order to answer both of these questions, the FDISO needs a keen understanding of safety concepts.

■ **Note**
An effective FDISO is well grounded in recognized safety concepts (theory), which gives that FDISO "uncommon sense."

To understand safety concepts, the FDISO must look at the components that make up the operational environment: procedures, equipment, & personnel (Figure 2-1). In order to make the operational environment safer, these components should be evaluated with regard to safety and then gauged with solid risk management concepts. Sounds simple? This chapter discusses procedures, equipment, and personnel and ends with risk management, collectively giving the FDISO a solid safety concept foundation.

SAFETY IN THE OPERATIONAL ENVIRONMENT

Procedures

As stated previously, the three basic components form an operational environment like the one found at a working structure fire or multicar extrication. If the operation is to be safe, each component needs to be addressed.

Procedures or processes (Figure 2-2) are the structure from which all activity at an incident begins. A first-arriving engine at a fire alarm activation probably follows a set of procedures or start a series of processes to investigate the

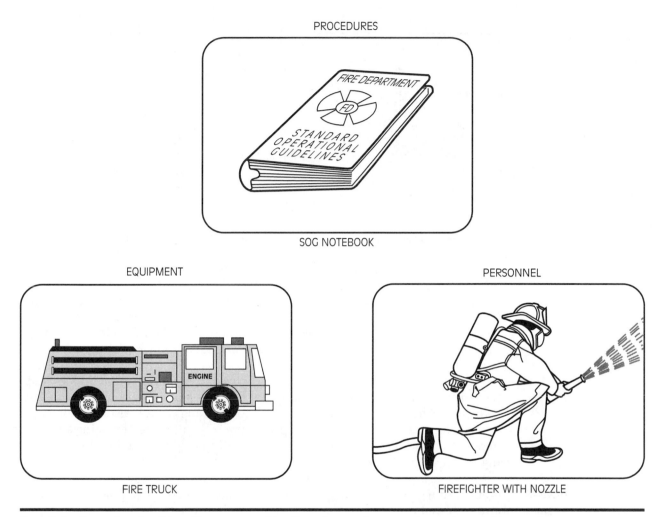

PROCEDURES

SOG NOTEBOOK

EQUIPMENT

FIRE TRUCK

PERSONNEL

FIREFIGHTER WITH NOZZLE

Figure 2-1 *Operational safety relies on a combination of procedures, equipment, and personnel.*

alarm. Procedures can be classified as formal or informal. Formal procedures are those that are in writing, either as a Standard Operating Procedure (SOP) or Standard Operating Guideline (SOG). In some departments, formal procedures are derived from standard evolutions or lesson plans. These evolutions and lessons can be drilled periodically, on a rotating basis, to ensure that a crew's response to a given situation is appropriate. The key idea here is that the procedures and evolutions are in writing. Some departments adopt training manuals as the department's operating standard. In cases where a manual gives choices, for instance a hose load, the department can circle the chosen load in the manual. In taking this approach, the department achieves consistency in its operations.

PROCEDURES

SOG NOTEBOOK

Figure 2-2
Procedures.

Many departments around the country have adopted the SOG instead of SOP in recognition of the belief that a guideline is more flexible and therefore more usable by line officers and incident commanders. A recommendation to use this vernacular was made to the Lisle-Woodridge Fire Protection District (Illinois) by its insurance carrier as part of a scheduled audit. The Poudre Fire Authority (Fort Collins, Colorado) uses General Operating Guidelines or as they call it, "goggles," to outline procedures for fire companies. The Loveland, Colorado, Fire and Rescue Department uses Emergency Operation Policies and General Operations Policies (EOs and GOs).

Informal procedures are those processes and operations that are obviously part of the routine of a given department but may not be written. Informal procedures are passed on through new member training as well as day-to-day operations. One example of an informal procedure is the practice of placing a full-face hood across bunker boots so that the firefighter must don the hood before placing his or her feet in the boots. Another example is the apparatus operator who uses a grease pencil to make status marks on pump panel gauges as a quick reference. At a glance the operator can see if changes have taken place. Both formal and informal procedures can guide the overall safety of a department.

The obvious first step in the development of any SOP or SOG (we use SOG for the remainder of this chapter) is to develop an administrative process to create, edit, alter, or delete established processes. Once the process is in place, a general format for SOG appearance and indexing is necessary. As you can see in Figure 2-3, the department has chosen to classify SOGs by topical areas. Figure 2-4 shows a typical SOG format.

Once topic areas have been defined, writing the SOGs can begin. Which topic is most important? The department can approach this one of two ways, either of which can be effective. One way is to perform a needs assessment and flag the areas where line firefighters and officers need guidance. The other way is to look at external influences such as OSHA regulations, Insurance Services Office Rating schedules, NFPA standards, and such, and determine which areas would impact the department most by *not* having an SOG. Departments choosing this

■ Note
Both formal and informal procedures can guide the overall safety of a department.

1. 1.1 Incident Command System
2. Emergency Ground Operations
   2.1 Rapid action group*
   2.2 Gas/odor investigation*
   2.3 Auto alarms*
   2.4 Train fires*
   2.5 Vehicle fire*
   2.6 Fires at postal facilities*
   2.7 Emergent driving procedure
   2.8 Kaneb pipeline response
   2.9 Volunteer and fire apparatus placement for motor vehicle accidents (MVAs)
   2.10 Operations involving thompson valley ambulance
   2.11 Minimal staffing for interior firefighting*
   2.12 Fire ground formation and activation of companies*
   2.13 Standard fire attack procedures/dwelling fires*
3. Alarm Levels/Dispatching
   3.1 City alarm level assignments
   3.2 Rural alarm level assignments
   3.3 Fire resource officer*
   3.4 Fire alarm panel operation and response policy
   3.5 Mutual/automatic aid agreement*
   3.6 Staffing considerations during adverse weather conditions
   3.7 Cancellation procedures for emergency medical service (EMS) and MVA incidents
4. Hazardous Materials
   4.1 Hazardous materials operations*
5. Emergency Medical Services
   5.1 Duties for non-EMS certified personnel
6. Aircraft Rescue and Firefighting (ARFF)
   6.1 ARFF standby policy
7. Technical Rescue and Special Operations
   7.1 Vehicle extrication
   7.2 Rope
   7.3 Trench*
   7.4 Collapse rescue*
   7.5 Confined space*
   7.6 Farm equipment and industrial rescue*
   7.7 Loveland dive rescue standard operating procedures
   7.8 Use of Civil Air Patrol

*These policies still need to be developed and/or approved.

Figure 2-3 *Sample SOG index.*

Purpose:

To establish policy and direction to all department members regarding minimal staffing and resource allocation for safe and aggressive interior structural firefighting.

Responsibility:

It is the responsibility of all officers and firefighters engaged in firefighting operations to adhere to this policy. The Incident Commander is accountable for procedure included within this policy.

Procedure:

1. This policy is applicable to situations where the Incident Commander (IC) has made a tactical decision to initiate an *offensive fire attack*, by firefighters, inside the structure. Additionally, tactical firefighting assignments that expose firefighters to an atmosphere that is *immediately dangerous to life and health* (IDLH) dictate the application of this policy.[1]

2. Prior to initiating interior fire attack or exposure of firefighters to an IDLH atmosphere, a ***minimum of four (4) firefighters shall assemble on scene***.[2] These four members shall utilize a "two-in, two-out" concept.

3. The *"two-in"* firefighters that enter the IDLH atmosphere shall remain as partners in close proximity to each other, generally fulfilling the operational role as the *FIRE ATTACK GROUP*. As a minimum, the *"two-in"* firefighters entering the IDLH atmosphere shall have full PPE, with SCBA and PASS device engaged, and have among them a two-way portable radio, forcible entry tool, and flashlight or lantern.

[1] An *IDLH atmosphere* can be defined as an atmosphere that would cause immediate health risks to a person who did not have *Personal Protective Equipment (PPE)* and/or *Self-Contained Breathing Apparatus (SCBA)*. This includes smoke, fire gases, oxygen deficient atmospheres, or hazardous materials environments. For Loveland Fire and Rescue application, an IDLH atmosphere can be further defined as an environment that is *suspected* to be IDLH, has been *confirmed* to be IDLH, or *may rapidly become* IDLH. The use of full protective equipment including an activated SCBA and an armed PASS device is mandatory for anyone working in or near an IDLH atmosphere.

[2] The firefighters must be SCBA qualified and capable of operating inside fire buildings without immediate supervision.

Figure 2-4 *Sample SOG format.*

route find that items such as personal protective equipment (PPE), self-contained breathing apparatus (SCBA), equipment maintenance, and patient care get high priority in the writing effort. As a starting place, SOGs should exist for:

- Use of PPE and SCBA
- Care and maintenance of PPE and SCBA
- Emergent (lights, sirens) driving

- Apparatus maintenance
- Accident and injury procedures and reporting
- Incident scene accountability
- Emergency evacuation at incidents
- Use of the incident command system
- EMS standard of care
- Infection control
- Employee right-to-know (hazards of firefighting)

What makes a good SOG? The answer is simple—your firefighters follow it! This is easier said than done. Good SOGs start with good writing. Good writing starts with a clear outline and use of simple language.

The outline can come from an officer's meeting, direction from the chief, or a sample from another department. Using the format in Figure 2-4, the author should address the reason (purpose) for the SOG followed by the responsibility each affected member has to the SOG. Some SOGs have responsibilities at different levels. For example: A firefighter may have the responsibility to ensure his or her accountability tag is checked into a staging board. The company officer, on the other hand, may have a responsibility to oversee that this is done as well as a responsibility to process accountability tags based on his or her company assignment. The department would have responsibility to make sure a usable policy exists for accountability, as well as making sure that training on the system is provided and that each firefighter has a tag. Some qualities of a good SOG include:

- Simple language
- Clear direction
- Tested technique
- Easy interpretation
- Applicability to many scenarios
- Specific only on critical or life-endangering points

Safety
A well-applied SOG improves departmental safety!

Note
The FDISO's role in procedures deals with application and review, rather like a quality control officer.

The benefits of clear, concise, and practiced SOGs are numerous. SOGs can become a training outline, a tool to minimize liability, and certainly a tool to guide your members. In any case, a well-applied SOG improves departmental safety!

The FDISO's role in procedures deals with application and review, rather like a quality control officer. To be effective, the FDISO needs to know which SOGs are being applied to a given situation and whether that SOG is being accomplished as intended. In cases where the SOG is not being used appropriately, or in the case where an SOG exists but is not being used, the FDISO needs to interpret whether the actions of firefighters meet the intent of the SOG or if an injury potential exists because the SOG is not being followed. This application practice puts the FDISO in the *best* place to suggest changes to SOGs or even help create new SOGs for the

■ **Note**

The FDISO who wit-
nesses a failure to
follow SOGs should
make a notation during
an incident and ensure
that this subject is
brought up during
postincident analysis or
during the next sched-
uled safety committee
meeting.

department. The FDISO who witnesses a failure to follow SOGs should make a notation during an incident and ensure that this subject is brought up during postincident analysis or during the next scheduled safety committee meeting. If the failure to follow a SOG presents a situation that is potentially or imminently dangerous, the FDISO must intervene in order to prevent an injury.

Equipment

In the past few years the fire service has seen a venerable explosion in new equipment (Figure 2-5) designed uniquely for improved safety. What works? What does not? What is a fad? What is here to stay? What is fluff? What is essential? How much does it cost? Is it worth it? How long will it last? Will it be outdated soon?

So many questions, and so much time spent answering these questions. Too often, fire department efforts to improve firefighter safety are focused on equipment. Compound this by a tendency to blame equipment following an accident: It is much easier to blame equipment than it is to blame a person. To some degree, this blame is predictable. We call this the Blame Game:

Blame Game 1

"Chief, it wasn't my fault, the darn _____ broke."

or

"Chief, if we only had one of those new _____, this never would have happened."

■ **Note**

Equipment helps, but it is arguably the least important factor in the operational triad of procedure, equipment, and personnel.

Equipment helps, but it is arguably the least important factor in the operational triad of procedure, equipment, and personnel. In looking at building a

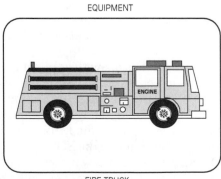

Figure 2-5
Equipment.

safety concept understanding, let us see how equipment can improve a department's safety. The following factors can be used to evaluate equipment and its selection and use.

Department Mission By looking at a fire department's scope of offered services, one can quickly and easily determine if equipment necessary for safe operations is missing. To start, department officers should meet and make a list of the types of incidents that are handled by their jurisdiction. They should then make a corresponding list of equipment necessary to safely handle that incident to the degree that the fire department is responsible. As an example, many departments faced an influx of service calls for the activation of carbon monoxide (CO) detectors. These devices were designed to activate with as little as 20 parts per million of CO present in air. Many fire and rescue agencies lacked calibrated instruments to confirm the presence of CO in a home. From this national experience, many departments began carrying high-tech, multigas monitors to assist in the safe handling of these types of incidents.

Next, officers must discuss the equipment list and place a check mark next to those items that are *essential* to safe operation and a circle next to those nice-to-have items. It is important to stay focused in this process. Although policy and procedure, as well as training needs, are important components here, they may distract the process. The officers should now compare this list with the equipment already on hand. Items that need to be obtained can then be prioritized for budgeting and appropriation.

External Influences When looking for equipment to make incident operations safer, one need only look to the advertising pages of the many trade journals or scan through the dozens of safety supply catalogs sent to the firehouse. A better start would be to look at equipment that may be *required*. Although requirements vary from state to state, you can look to the following for help on what may be required equipment:

• Occupational Safety and Health Administration (OSHA) Regulations. Known as "CFRs" (Code of Federal Regulations), these codes often outline the equipment required for a given process to be accomplished. At this writing, states covered under a state-sponsored OSHA plan may have more stringent equipment requirements for public agencies. Those without a plan do not require OSHA compliance from public agencies. For example, the state of Colorado has no state plan—fire departments have no obligation to follow OSHA. Washington state, however, has its own plan (Department of Labor and Industries) and compliance is mandatory for all public agencies. OSHA reform is being debated at the federal level as of this writing. Soon all public agencies may fall under more restrictive federal OSHA regulations.

• National Fire Protection Association (NFPA) Standards. The vast majority of fire service equipment is tailored to meet or exceed NFPA standards. These consensus standards are designed to offer a minimum acceptable standard for equipment design, application, and maintenance.

• National Institute for Occupational Safety and Health (NIOSH), American National Standards Institute (ANSI), and Underwriters' Laboratories (UL). Many equipment manufacturers use these agencies to show that their product meets or exceeds design and performance requirements for a given type of equipment.

Equipment Maintenance As most firefighters know, equipment utilized for incident operations is no better than the care and maintenance it receives. Following an accident, much time is spent evaluating the performance of a piece of equipment. Often, the piece of equipment was found to be inappropriate for the use or it was operationally unsound.

Because many different firefighters may use and maintain a given piece of equipment, complete documentation of repairs and maintenance is essential. Further, a complete set of guidelines should be developed or adopted for care of essential equipment. Rich Duffy, Director of Occupational Health and Safety for the International Association of Firefighters (IAFF), and Chuck Soros, retired Chief of Safety for Seattle, Washington, suggest seven items to be considered when writing equipment guidelines:[1]

1. Selection
2. Use
3. Cleaning and decontamination
4. Storage
5. Inspection
6. Repairs
7. Criteria for retirement

The Right Equipment A quick look at firefighter injury and death statistics will show *what* equipment can make a difference. The following are equipment items that have made a difference in firefighter safety over the past few years. This list is designed to stimulate conversation in your department, hopefully leading to wise equipment changes or purchases. As with any piece of equipment, it is worthless if not used and maintained by trained firefighters.

Personal Protective Equipment (Figure 2-6)
Accountability tags/Passports
Disposable EMS masks/gloves
Water-free hand disinfectant
High visibility materials/colors
Nomex/PBI/P84 materials

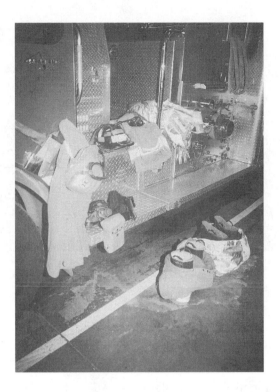

Figure 2-6
Firefighters must now choose from multiple types of protective clothing.

Lightweight, positive-pressure SCBA with nonmelting straps

Integrated personal alert system (PAS) devices

Apparatus (Figure 2-7)

Enclosed cabs

Intercom/radio headsets

Three-point seat belts

Intersection strobes

Quick-deploy scene lighting

800MHz radios/trunk systems/mobile data terminals

Cellular phones with priority override

Laptop computers

Vertical exhaust pipes

Wide reflective trim

Roll-up compartment doors and roll-out trays

Global positioning systems (GPS)

Automatic vehicle locators (AVL)

Figure 2-7 *Modern fire apparatus are continuing to improve to include more safety for firefighters.*

Tools (Figure 2-8)

Multigas detectors/monitors

Speed shores

Rehabilitation kits

Command/status boards

Position identification vests

Disposable airway management adjuncts

Two-way radios for each firefighter on a crew

Infrared video/visual cameras

Figure 2-8 *"High-tech" tools allow firefighters to work safer and monitor our health.*

Figure 2-9 *Physical fitness equipment is actually "firefighter safety" equipment.*

Station Equipment (Figure 2-9)

Exhaust removal systems

Aerobic exercise devices

Dedicated disinfection systems/areas

Fire suppression sprinkler systems

Monitored fire and smoke alarm systems

Extractors for washing structural firefighter clothing

Open-air/forced-air protective gear storage systems

The effective FDISO understands the relationship of equipment to safety. Ted Jarboe, Deputy Chief of Management Services for Montgomery County (Maryland) Fire and Rescue Services said it best when he said, "Don't let technology lead you into taking more risks."[2]

!Safety
"Don't let technology lead you into taking more risks."

Personnel

When discussing people as a safety issue, many opinions, philosophies, and emotions are considered. It is easy to blame a safety deficiency on equipment or on the poor or nonexistent procedure. Once again, we can see this in our day-to-day station dialogue:

Blame Game 2

"Chief, if we only had a procedure that said _____, this would never have happened."

PERSONNEL

FIREFIGHTER WITH NOZZLE

Figure 2-10
Personnel.

or

"Chief, I'm really sorry Firefighter _____ got hurt. But I did it just the way SOG # _____ suggested."

It is more difficult to address the people component of the safety triad because of the opinions and emotions involved. Regardless, a method of improving safety by addressing people is essential. Three factors contribute to a person's ability to act safely: training, health, attitude (see figure 2-10).

Training A successful safety program is usually synonymous with a successful training program. Likewise, an organization plagued by injuries or suffering from costly accidents usually has a deficiency in its training effort. As it relates to safety, what makes a training program effective? First, some specific qualities should be present in the training:

■ **Note**
A successful safety program is usually synonymous with a successful training program.

- Clear objectives
- Applicability to incident handling
- Established proficiency level
- Identification of potential hazards
- Definition of the acceptable risk to be taken
- List of options should something go wrong

Second, the training program must include the right subjects. Arguments can be made on which training subjects or behaviors are most important to safe operations. A common list can be created, however, based on firefighter injury and death statistics. Figure 2-11 is a list of training subjects that directly affect incident safety: If these subjects are practiced and appropriately applied, incident operations will become more safe. An expectation of the depth of understanding and methods to achieve this are also suggested.

Essential Training Subjects for Increased Incident Safety	
Subject	**Degree of Understanding**
• Personal protective equipment	Mastery
• Accountability systems	Mastery
• Company formation and team continuity	Mastery
• Fire behavior and phenomena	Proficient
• Incident command systems	Proficient
• Apparatus driving	Proficient under stress
• Fitness and rehabilitation	Practitioner

Figure 2-11 *Injury and death statistics suggest that essential training subjects be addressed.*

Health The safety and well-being of any firefighter increases with the health of the individual firefighter. Much has been written on the benefits of healthy firefighters, most of which centers on *physical* health. Stress continues to lead in causes of firefighter duty deaths and is a significant contributor in injuries. To handle the inherent stress of fire fighting, each firefighter's body must be accustomed to, and capable of handling stress. Additionally, the firefighter must protect him- or herself from, and prevent the spread of, communicable diseases and infections. Keys to improving physical health, and therefore department safety, include:

> **Safety**
> Stress continues to lead in causes of firefighter duty deaths and is a significant contributor in injuries.

- Annual health screening for all firefighters and line officers
- Vaccination and immunization offerings
- Employee Assistance Programs (EAP) for alcohol and drug dependencies as well as workplace stress
- Process to determine Fit for Firefighting
- Work hardening and mandatory ongoing fitness programs
- Firefighter fueling (nutrition) education

Attention to physical health is indeed important. Mental health is also important to firefighter safety. Thanks to the recent attention to critical incident stress, fire departments are becoming more concerned with the mental health of firefighters. Keys to supporting the mental health of firefighters include:

- Training and understanding of critical incident stress signs and symptoms
- Creating a process to have a critical incident stress debriefing (CISD) team activated or available for unusual events or at the request of one or more responders
- Including the firefighter's family in department events

- Making available an Employee Assistance Program (EAP) for job stress coaching, work/family issues

Attitude Of all the people factors affecting safety, attitude is the hardest to address. Perhaps this is why firefighter attitudes receive the least amount of attention when dealing with safety. As noted above, people tend to compound safety problems by trying to place blame after an accident.

Blame Game 3

"Chief, it's not my fault. How was I to know _____ was going to _____."

or:

"Chief, I was just doing what I was told. I didn't think _____ could ever happen to us."

Many factors affect the attitude of a given individual, not to mention the fact that attitudes are dynamic. Of the many factors that affect safety attitudes, the following represent a few prevalent factors in the fire service:

The Department's Safety Culture The culture of an organization is the ideas, skills, and customs that are passed through generations. How does one see and, more importantly, measure a department's safety culture? One way to illustrate a safety culture is to provide two conversation examples from the fire house:

SOMETOWN FIRE STATION 1:

APPARATUS OPERATOR: "Hey Lieu', I'm not going to put the cover back on the grease pit, Jim's bringing down Engine 2 a little later."

LIEUTENANT: "Hope none of the visitors fall in."

APPARATUS OPERATOR: "Well, there's a big yellow stripe around it."

LIEUTENANT: "OK, but I didn't see anything."

ANYTOWN FIRE STATION 1:

APPARATUS OPERATOR: "Hey Lieu', I'm not going to put the cover back on the grease pit, Jim's bringing Engine 2 down a little later."

LIEUTENANT: "That may not be a good idea, the station is wide open and I'd hate to see anyone fall or trip into that pit."

APPARATUS OPERATOR: "Well, there's a big yellow stripe around it."

LIEUTENANT: "I know, and I know the cover is heavy and hard to move back and forth...But I'm serious, I don't want anyone to run into that hole - why don't you put some traffic cones around it?"

APPARATUS OPERATOR: "Good Idea." <walks away>

It is easy to see the two different attitudes toward safety. The culture of the department may be reflected in daily conversations or in actual actions taken by the department. In a unique example, the Denver Fire Department experienced a significant accident where two apparatus collided with each other enroute to a reported fire. The department ruled that the most significant factor leading to the accident was the department's attitude that condoned competition between companies to be the first one to get water on the fire.

The Department's Firefighter Death or Injury History A firefighter duty death will often shock a department's members into an attitude change. Dr. Morris Massey calls this a "significant emotional event" in his renowned video tape, *You Are What You Were When.* A traumatic death is capable of changing a person's "value programming." Most often, this change is one toward a more healthy safety attitude. Although some departments have dismissed a death as a pure accident, most seek to change the way they do business to ensure that the event never repeats itself.

The Example (or Lack of) Set by Line Officers and Veteran Firefighters Legendary Notre Dame football coach Knute Rockne once said, "One man practicing sportsmanship is far better than a hundred teaching it." The same can be said about safety. Is safety merely being taught? Or is it being practiced? One look at your own department can show if the following safety items, like Knute Rockne's sportsmanship, are being practiced:

- *Crews or company members are not only watching themselves, but watching their team members.* Some examples may include crews that make quick, head-to-toe checks of each other just prior to interior firefighting entry; crew integrity that prevails at ALL incidents, ALL the time; company officers who give brief safety reminders prior to tactical assignments; tool operators who voluntarily pass a tool to another operator when initial efforts are unsuccessful; firefighters give protective equipment reminders that are welcome and expected; REHAB and SCBA attendants who are organized for quick recognition of fatigue and equipment problems.

- *Work areas are neat and organized.* A Pennsylvania safety officer once said that he could tell if a department had embraced safety from a simple tour of the apparatus bay and the firehouse lounge. Although the evidence is clear—a clean workplace is a safe workplace—it is best to look to the actions of individuals. Do firefighters routinely correct trip hazards while working on a project? Are compartment doors closed as soon as a tool is retrieved? Are doorways kept clear at the station as well as the incident? Do apparatus operators routinely point out obstacles to masked-up firefighters? Does out-of-service equipment get immediate flagging at the incident scene?

- *Drivers are calm, consistent, and attentive.* Safe drivers are usually the ones who follow a simple routine that begins with a confirmation of the incident location. The confirmation is followed by communication with the company officer about location, route, or traffic concern. The driver proceeds to the apparatus in such a way as to get a 360° or three-side view of the apparatus. The driver does start up and belt check, then a passenger check. After a GO signal, he or she then makes a mirror check, looks up at the bay door, then a visual of the apron. Finally the vehicle moves. The driver stops before entering the roadway. Out on the road, you would hope to get the sense of control with very few quick-jerk movements. Acceleration is smooth, braking is firm and straight, cornering is like riding a rail. The driver's eyes are always moving and attentive. Face muscles are relaxed and both hands are graceful in steering and shifting.

■ **Note**
Attitude changes are slow, often emotional, and require lots of buy-in.

- *Observations are openly shared.* As an incident safety officer, one of the most reassuring measures of instilled safety values come when firefighting teams and company officers report hazards to *you.* If this is happening, it will be coupled with your observation that functioning personnel are spending more time looking UP and looking AROUND. Teams are pointing at walls, wires, and windows. You'll hear more "Watch out for this . . ." or "keep an eye on that . . ." among the crews. Crews, themselves, will put up barrier tape around firefighter hazards or collapse zones. The more you see and hear these behaviors, the further advanced the safety values of the firefighters.

Hopefully you can look at the foregoing list and see where your department stands. Remember, however, that attitude changes are slow, often emotional, and require lots of buy-in. Set personal goals for yourself—be the example—and then try working small but steady changes into the department. Be patient.

RISK MANAGEMENT

risk management
the process of minimizing the chance, degree, or probability of damage, loss, or injury

Every day we, as individuals, take risks. Risks can simply be defined as the chance of damage, injury, or loss. **Risk management** is the process of minimizing the chance, degree, or probability of damage, loss, or injury. Most risk managers use a five-step risk management process called Classic Risk Management. Classic Risk Management (Figure 2-12) was the risk management method chosen by the National Fire Academy for its Health and Safety Officer course. An understanding of this process can help the incident safety officer make a difference.

Five-Step Risk Management

Step 1: Hazard Identification In essence, hazard identification is the primary function of an incident safety officer. This book explores how to identify hazards through an FDISO Action Model covered in Section 2. Important to mention here,

Five-Step Risk Management

1. Identify hazards

2. Evaluate hazards

3. Prioritize hazards

4. Control hazards

5. Monitor hazards

Figure 2-12   *The five-step Risk Management Model is used by risk managers worldwide.*

however, is the fact that many operations that are routine to our profession and may not be viewed as hazardous by us, but are considered dangerous to most outsiders. The test to see if an operation or environment is hazardous can come from a review of injury and accident data. To find this data, one need only look to their local risk manager or, in some cases, the state worker's compensation office.

Step 2: Hazard Evaluation   In this step, a value is established for a given hazard. First, however, the hazard needs to be viewed in terms of *frequency* and *severity*. Frequency is the probability that an injurious event can happen and can best be described as low, moderate, or high based on the number of times that a particular hazard is present or the number of times an injury resulted from the hazard. The same descriptions can be applied to severity. Severity can be viewed as a harmful consequence or cost associated with injury or damage from a given hazard. With this approach, one can see that any given hazard falls into one of nine categories as indicated in Figure 2-13.

Hazard Evaluation Matrix

		Frequency		
		High	Moderate	Low
S e v e r i t y	**High**	High/High	High/Moderate	High/Low
	Moderate	Moderate/High	Moderate/Moderate	Moderate/Low
	Low	Low/High	Low/Moderate	Low/Low

Figure 2-13   *A recognized hazard should be placed in one of these boxes based on the potential severity and frequency of that type of hazard.*

With this matrix, a value can be added to a given hazard to help determine the priority, or level of importance, the given hazard should receive. This leads us to the next step.

Step 3: Prioritize Hazards Obviously, a hazard that ranks as high frequency/high severity is one we want to avoid at all cost. Conversely, a low frequency/low severity hazard is one that does not warrant much time or effort in correcting. A good example here is the classic division of fireground strategies: offensive and defensive. A well-involved fire that has captured the ceiling space is a high frequency/high severity situation, that is, one that will produce a devastating injury in virtually every case. On the other hand, we do not spend much time developing procedures for a coffee pot overheat. One method to simplify this matrix and priority system is to divide the matrix into three hazard classes as indicated in Figure 2-14: Priority one, two, and three.

As a starting place, the FDISO should address any hazard that falls in the priority 1 category. During some incidents, the FDISO may never get an opportunity to address priority 3 items. If the incident is such that only priority 1 hazards get attention, then the FDISO or Incident Commander may consider expanding the safety role to include more FDISOs.

Step 4: Control Hazard Once the hazard has been prioritized, efforts can begin to minimize exposure to the hazard. This effort can range from total avoidance, to hazard transfer, to hazard adaptation. For the FDISO, hazard avoidance and

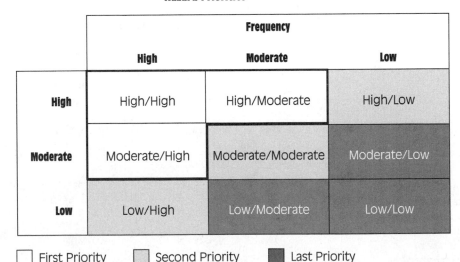

Hazard Priorities

		Frequency	
	High	**Moderate**	**Low**
High	High/High	High/Moderate	High/Low
Moderate	Moderate/High	Moderate/Moderate	Moderate/Low
Low	Low/High	Low/Moderate	Low/Low

☐ First Priority   ☐ Second Priority   ■ Last Priority

Figure 2-14 *Once a hazard is classified in one of these boxes, a priority can be assigned to the hazard. This helps the FDISO juggle multiple hazards.*

mitigation

actions taken to eliminate a hazard or make the hazard less severe or less likely to cause harm

transfer are not always possible. Hazard adaptation, however, is the control method most often employed on scene. Hazard adaptation can be accomplished many ways and in many forms. All adaptation methods are designed to make the hazard less severe for the exposed firefighter—a process called *mitigation.* Mitigation is usually accomplished through ten classic countermeasures:

1. Prevent the creation of the hazard.
2. Reduce the extent of the hazard.
3. Prevent the release of the hazard.
4. Modify the rate of release of the hazard.
5. Separate the hazard by time and space.
6. Separate the hazard by a barrier.
7. Modify the basic quality of the hazard.
8. Make the hazard resistant to injury.
9. Counter the damage done by the hazard.
10. Stabilize/repair the damage done by the hazard.

Step 5: Monitor Hazards If the risk management approach is effective, the department should see a decline in injuries, accidents, and close calls. Changes in equipment, staffing, procedures, and so forth can create, alter, or eliminate hazards. Constant monitoring can catch these changes and, hopefully, lead to proactive hazard control. In one example, a city experienced an increase in back injuries. The city risk manager hired a noted back injury prevention specialist, and, in less than 2 years, virtually eliminated all back injuries. The program paid for itself in those 2 years by a reduction in worker's compensation claims. This, in turn, lowered the city's loss history, which lowered its annual premium.

At an incident, the FDISO is always monitoring hazards, even after hazard countermeasures have been implemented. This monitoring creates cyclic thinking, that is, the ability to revisit hazards and continually weigh the operations and the environment to see if a hazard is truly being mitigated. Just as a fire is dynamic, so must be the FDISO in his or her evaluation of risk. An excellent phrase that captures the essence of this last step of hazard monitoring was presented in the U.S. Fire Administration's publication, *Risk Management Practices in the Fire Service:* "Risk management is a system, not a solution."[3]

This chapter has provided a basic foundation of safety concepts. Look at your environment and see how these concepts can help you identify and address problems. The practiced FDISO can take any observation or hazard on the incident scene and immediately classify its priority, mitigation countermeasure, and its need for evaluation. This is the hallmark of making a difference.

Summary

To be effective, the incident safety officer must have a solid foundation in general safety concepts and risk management. The role of workplace procedures, equipment, and personnel should be understood by the FDISO as well as how to achieve a safe workplace through the evaluation and improvement of these areas. In many cases, improvement may be difficult, especially in the areas of attitudes. Risk management is the process of minimizing the chance for an injury or loss or minimizing the degree of that injury or loss. The most common approach to risk management is the five-step classic risk management model. The incident safety officer can apply the concepts of classic risk management and actually prioritize hazards that need attention. This hazard reduction or mitigation can be achieved through many typical countermeasures.

Review Questions

1. List the three elements that affect workplace safety.
2. Explain the difference between formal and informal procedures.
3. Describe four qualities of a well-written procedure.
4. List and describe the external influences that can affect safety equipment design and purchase.
5. List and briefly describe the three factors that influence a person's ability to act safely.
6. Define risk management.
7. List and explain the five steps of classic risk management,

Notes

1. R. Duffy and C. Soros, *The Safety Officer's Role* (Ashland, MA: Fire Department Safety Officers Association, 1994).
2. T. Jarboe, *Managing Operations,* Book 3, Company Officer Development Series (Ashland, MA: International Society of Fire Service Instructors, 1991).
3. Federal Emergency Management Agency, United States Fire Administration, *Risk Management Practices in the Fire Service,* FA-166, December 1996.

Chapter

3

Designing an Incident Safety Officer System

Learning Objectives

Upon completion of this chapter, you should be able to:

■ Discuss the reasoning for preplanning the response of an incident safety officer.

■ List four examples where an automatic FDISO response should take place.

■ List four examples of when an incident commander should automatically delegate the safety responsibility to an FDISO.

■ List and discuss the advantages and disadvantages of using various methods to ensure an FDISO arrives on scene.

■ Discuss the authorities suggested for incident safety officers by NFPA standards.

■ List several tools that help the FDISO be effective on scene.

The incident came in as a motor vehicle accident over a half-hour ago. Since then, you, the IC, have had to order two more engines, four ambulances, and a heavy rescue. Never did you expect to find a six vehicle pileup that included a hotel shuttle bus with ten hurt and trapped occupants at the center of the wreck. The rain is letting up but everything is soaked. The incoming units are just arriving and you've passed out assignments. The media is arriving. The highway patrol is pressuring you to reopen the blocked highway. A medical evac helicopter is due in about 5 minutes and another is needed. It dawns on you that you haven't delegated the safety officer position. Staging is empty and all the company officers are committed to critical tasks. You order another engine to come in hoping you're not too far behind the incident curve. With heavy hand, you write down on your incident worksheet and circle it several times: "We need a duty incident safety officer system!"

The design and implementation of a fire department incident safety officer (FDISO) program or system can make the difference in whether a program is effective. The program should address some key elements, including:

- Who responds and fills the FDISO role
- What type of incidents dictate the use of an FDISO
- What tools are necessary to assist the FDISO

Most incident commanders, fire chiefs, and firefighters would agree that incident safety officers are necessary and valuable at significant or complex incidents. The biggest obstacle, however, to an effective FDISO program falls in the design of the system to get the FDISO to the scene, because mostly we (the fire service) rely on the incident commander (IC) to make a decision sometime during an incident on whether a safety officer is necessary. For example, an emergency call is received and a programmed response is initiated. Once on scene, the incident command system is implemented and actions are taken to begin mitigation. If the incident is significant or overcomes the initial response, additional resources are requested. At about this time, an incident safety officer becomes a consideration for the incident commander. In this case, the IC is *reactive* in the delegation of firefighter safety duties. If an incident commander is truly going to make a difference at an incident scene, the delegation of the safety function needs to be *PROACTIVE*. To be proactive in the delegation and placement of the FDISO, the fire department needs to *PREPLAN* the FDISO response. This chapter explores *WHY* you need to *PREPLAN* an FDISO response as well as covering the *WHEN, WHERE,* and *HOW* of an effective FDISO system.

■ Note
If an incident commander is truly going to make a difference at an incident scene, the delegation of the safety function needs to be proactive.

PREPLANNING FDISO RESPONSE

Some incident commanders believe that any fire officer should be able to fill the FDISO position, at any time, under any circumstance, at the will and want of the incident commander, therefore, the agency really does not need to create an FDISO system. Not only is this thinking flawed, it is dangerous. Just as incident commanders have various levels of knowledge and expertise, so do other fire officers. Likewise, the requirements necessary to be a fire officer may change from department to department, a problem if mutual aid situations arise. Further, the emphasis placed on safety may vary from one IC to another IC. As stated in Chapter 1, firefighter death and injury statistics suggest that more can be done to improve firefighter safety. Those same statistics show that the majority of deaths and injuries on the fireground occur at residential structure fires.[1] Logic (as opposed to traditional fireground thinking) suggests that we send an FDISO to all residential fires. We know, however, that some incident commanders are thinking, "Residential fire . . . couple engines . . . a dozen people . . . we can handle that, we don't need a Safety Officer." The statistics show otherwise. Consider this: The Incident safety officer is most effective when he or she arrives early in the incident. A few pragmatic graphs (as opposed to scientific), based on a typical residential working fire, will illustrate this point.

> ■ **Note**
> The incident safety officer is most effective when he or she arrives early in the incident.

Graph 1: Environmental Change

As applied to a residential structure fire, "environmental change" means fire propagation, building degradation, and smoke and fire volume.

As you can see in the graph (Figure 3-1), rate of change is measured and rated over time. During a fire in a structure, there is actually a routine, or even a

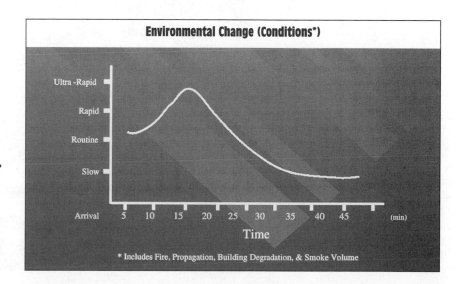

Figure 3-1 *The rapidly changing environment early in a fire is grounds for early appointment of a safety officer.*

rapid, rate of change upon arrival of fire crews. Granted, this depends on response time. In most large cities, fire stations are located so that a responding crew can intervene in a residential fire prior to flashover. Given this, the fire will be intensifying upon arrival (that is probably why someone called 911). At arrival (zero on the time line), the fire is starting to develop significantly. With that comes smoke production as more fuels become involved. Equally, the building itself is being attacked and thus becoming structurally degraded. If flashover happens prior to fire department control efforts, the environmental change becomes ultrarapid. Other events can cause ultrarapid change. Smoke explosions, backdrafts, partial collapse, and accelerant fuels can all cause ultrarapid change in the fire fighting environment. Once control efforts have begun, this rate of change can be stabilized and even reduced. It would seem to make sense to have an FDISO on hand to evaluate changes and find hazards during these periods of rapid and ultrarapid change.

Graph 2: Fireground Activity

Another area that suggests that an early FDISO assignment is advantageous is the amount of fireground activity. Let us look again at a typical residential fire. You can argue that upon arrival there are basically no tasks being performed (Figure 3-2).

Now, it is important to acknowledge that all responders are going through personal size-up and potential incident commanders are performing a lot of mental activity. Regardless, no actual crews are performing physical, active tasks in the dangerous environment upon arrival. As time on scene passes, the activities needed to control the incident increase: one to three task assignments, three to five task assignments. Within 20 minutes, the IC may have seven to ten simulta-

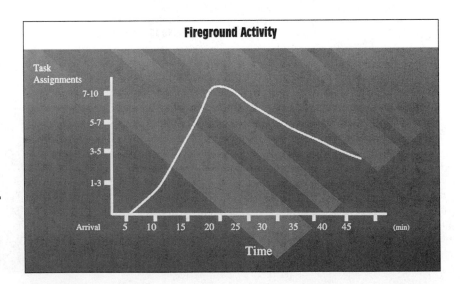

Figure 3-2 *The higher the number of task assignments, the greater the need for a safety officer.*

neous assignments orchestrated. These assignments may include search and rescue, exposure protection, ventilation, attack lines, backup lines, rapid intervention planning, and so forth. For some departments, the simple task of water supply is a significant chore that incorporates two or three task assignments that are going on simultaneously. The point here is simple: About 15 to 20 minutes into the incident, many tasks are going on concurrently. It should seem reasonable that a key to preventing injuries is to monitor the numerous activities that happen simultaneously in the first 20 minutes of the fire. The good Incident Commander wants an FDISO appointed early in the incident. Wouldn't it be great if you, as the IC, knew that an FDISO was on duty and responding prior to reaching the five to six task assignment operation?

!Safety

The good incident commander wants an FDISO appointed early in the incident.

Graph 3: Relative Danger to Firefighters

Chapter 2 outlined risk management. Practically speaking, every firefighter at a working fire performs some kind of risk management. With each assignment, the individual crews are responsible for a degree of risk management.

This mental activity translates into a relative danger for firefighters. Firefighters actually begin taking risks upon arrival, or at "zero time." At times, risks are taken just to determine the risk of the incident! An actual tragic example: The Denver Fire Department was called to assist the police department on a welfare check. A firefighter was assigned to climb a ladder to check for a possibly unlocked, upstairs window. He reached the top of the ladder and, without warning, was shot by the occupant of the house!

Throughout a fire some of our risk taking becomes extreme or high, usually right out of the gates (Figure 3-3). The perfect example of this immediate risk tak-

Figure 3-3

Firefighter risk taking is usually greatest in the early stages of an incident.

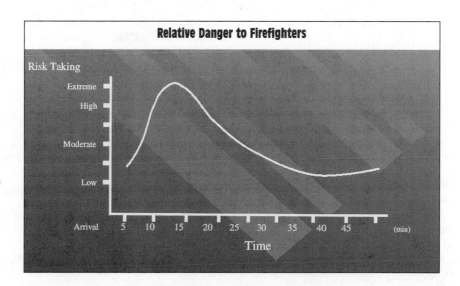

ing is search and rescue. A typical scenario is that a first-arriving crew starts setting up for an aggressive interior attack. They lay supply lines in, pull attack lines, get air packs on—all in preparation for an offensive attack. All of a sudden the incident commander comes from the back of the building after doing a 360° and yells that there is an immediate need for rescue. Three victims are hanging on the third floor balcony that need to be evacuated immediately. The crew drops the planned attack and begins taking risks to get to those people. As illustrated, it is early in an incident when risk taking is high. Volunteer, paid on-call, or pager-type fire departments who allow firefighters to respond directly to the incident scene may have an even greater degree of risk taking by the sheer nature of having firefighters on scene prior to fire fighting apparatus and equipment. Once again, the point is simple: Risks are usually greater early in an incident, therefore a safety officer is needed early.

Overlapping the Graphs

Fire and rescue departments that wish to reduce firefighter injuries and death will develop a system to get the FDISO on scene or appointed early in an incident. The overlapped graphs in Figure 3-4 clearly show that the first 20 minutes of an incident warrant monitoring of firefighters and firefighting operations. The early appointment of an FDISO gives another set of eyes, another viewpoint, another consultant for the incident commander. The only way to get an FDISO to the scene early is to predefine *when* an FDISO is required to be appointed and design a system to make sure a trained FDISO arrives on scene.

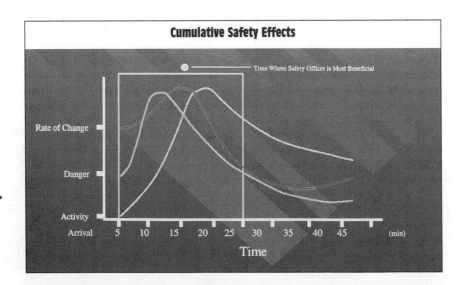

Figure 3-4   *The early assignment of an Incident Safety Officer is essential to firefighter safety.*

WHEN DOES THE FDISO RESPOND?

Earlier in this chapter we suggested that an incident safety officer be present on all working residential fires. Common sense suggests that an FDISO should be planned for highly technical or complex incidents. For these reasons, the response of an FDISO should be preplanned. This leads to proactive FDISO response. It is not suggested here that the incident commander's ability and authority to make decisions be taken away. What is being suggested, however, is that firefighter injury statistics show that we need to have an FDISO more often and sooner. If an IC can get an extra set of eyes on scene sooner—upon arrival at an incident—collectively, he will likely manage the scenes better and be more safe, which will reduce death and injury. In defining *when* an FDISO is required to respond, the individual fire and rescue department should develop guidelines for an automatic FDISO response as well as guidelines or circumstances that mandate automatic FDISO delegation during the incident.

Automatic FDISO Response

The goal of automatic FDISO response is to have a predesignated, trained FDISO programmed to respond to certain incidents that have a high risk to firefighters. Once again, the risk management concepts from Chapter 2 can be applied in determining these situations. Generally speaking, types of incidents that may require an automatic FDISO response include:

Residential or Commercial Structure Fires Upon receipt of a 911 or other report of a fire in a residential or commercial fire, an FDISO should be programmed into the response (Figure 3-5).

For some departments, dispatch personnel may need to add a level of screening here so that a "burnt toast" call is not classified as a structure fire. Some may argue that this would lead to an FDISO response to virtually every fire—something that a large department (because of cost) or small department (because of available staffing) may not find acceptable. Unfortunately, statistics do not justify this. Residential fires hurt and kill firefighters! Once we, as a fire service, force a significant decline in our injury problem, then maybe the idea of sending an FDISO to every structure fire is not necessary. Until then, its merits outweigh its cost. In the case of small or critically staffed fire agencies, an argument can be made that the fewer people that are available, the more important an FDISO becomes. In essence, each person on scene is taking a greater risk than with an appropriately staffed response.

I-zone
wildland-urban interface; where homes and businesses blend with hill or mountain areas

Wildland Interface Fires Fires in the wildland/urban interface (**I-zone**) can cause unique problems and present unusual choices for first responders. First-arriving company officers have to evaluate fuels, weather, topography, fire conditions, access, and defensibility of threatened structures (Figure 3-6).

Figure 3-5   *The report of an actual hostile fire should trigger an FDISO response.*

Figure 3-6   *Fires in the wildland or "I-zone" should trigger an automatic FDISO response.*

This combination leads to an environment well suited for the partnership of an incident commander and incident safety officer early in the incident. In some jurisdictions, the time of year and local fuel conditions may indicate the need for an automatic FDISO response. For example, when fire danger is rated "High" or "Extreme" by the forest service, the local fire department automatically sends an FDISO to any reported fire.

Specialty Team Incidents   Many fire and rescue departments have specific-function teams designed to handle the array of incidents outside the scope of standard engine and truck company functions (Figure 3-7). Examples include HazMat, dive rescue, heavy (urban) rescue, rope rescue, and wildland "hot shot" teams. Often, the activation of these teams takes a separate request or page. An FDISO should be included with this activation. Some of these teams include an FDISO specially trained to interface with the needs of the particular team. The now-famous **USAR teams** around the nation have specially trained safety officers that travel with the team. If the team does not include its own FDISO, a basic-trained FDISO can function as a common sense consultant to the team, support personnel, and incident commander.

USAR teams
urban search and rescue teams; highly trained and specially equipped to respond to significant structural collapses

Figure 3-7
Activation of specialty teams should also activate FDISO response.

Target Hazard Incidents In nearly every city, local firefighters can name a building where they hope they never get a fire "cause it'll be a killer." One advantage of computer-aided dispatch (CAD) systems is the ability to flag locations or occupancies that present unique hazards to firefighters. Examples may include chemical plants, historical buildings, stadiums, underground structures, and difficult or limited access occupancies. If an incident is reported at these locations, an FDISO notification process can be preprogrammed into CAD to initiate rapid response.

Aircraft Incidents The potential for high life loss, mass fire, and hazardous materials release are prevalent in aircraft incidents. Hence, the necessity that an FDISO be included with this type of response (Figure 3-8).

Departments that have airfield aircraft rescue and firefighting (ARFF) responsibilities have specialized equipment and procedures to minimize the risks associated with the aforementioned. An ALERT 3 (crash) can stretch these safety measures, especially if the airport has an ALERT 3 with an out-of-index aircraft. In Fresno, California, a military contract Lear jet crash-landed on a street in the middle of town and instantaneously started an aircraft, apartment, and commercial building fire as well as a HazMat and collapse incident. An incident commander would surely welcome the early arrival of a designated safety officer on this type of incident.

Figure 3-8 *Aircraft incidents can be complex, creating unsafe conditions beyond standard firefighting training.*

Weather Extremes Subfreezing temperatures or high-humidity, high-temperature situations present additional hazards to firefighters (Figure 3-9). Changes in strategies, tactics, and action plans may be required just because of the weather situation. If firefighters are unaccustomed to working in a given weather extreme, injury risk rises. For example, for the Colorado front-range (Denver, Colorado Springs), temperatures below 0°F and above 100°F happen only a few days a year, therefore firefighters may not be acclimated to such temperature extremes. In this case, accelerated rehabilitation and monitoring of firefighters is a solid argument for dispatching an FDISO. Additionally, snow depths greater than 3 feet are unusual for the same area. It makes sense to get an FDISO on scene early to help monitor the risks associated with the added stress of excessive snow. Granted, colleagues in parts of Minnesota or Alaska may think it silly to respond an FDISO just because the snow is a little deep. Consider also that snow accumulation on a rooftop of an assembly building in Dallas may just push it past its designed load capabilities. Acclimation is the key here: If the weather becomes extreme as defined by *your* jurisdiction, then consider dispatching an FDISO to all multi-company or working incidents.

Automatic FDISO Delegation

"Plan for the worst" is a code that most responding fire officers live by. The odds favor that the fire officer will have enough resources to handle the situation. Occasionally, however, the fire officer gets caught short, or worse, he or she thinks the situation has been handled only to have "Murphy" show up. In these

Figure 3-9 *Incidents during extreme weather should include the response of the FDISO.*

cases, it pays to have a predefined set of circumstances that lead to automatic delegation of the FDISO. By having these situations predefined, the incident commander is sure to get the valuable assistance that an FDISO can provide for escalating or multiple risk situations. Some examples:

Working Incident Generally speaking, a "working" incident can be defined as one where the first-due, on-scene resources are 100% committed and more resources are needed. In these cases, the control effort leads to greater risk taking by personnel (Figure 3-10). Here, the IC would call for additional assistance, which is an indicator that an FDISO is needed. Although language varies between jurisdictions, the striking of a second alarm can serve as the impetus for an automatic FDISO delegation.

Span of Control Exceeds Three Most incident command systems (ICS) recognize a span of control of five or less for emergency operations; that is, five or less crew leaders or functional assignments directly reporting to the IC. Those well versed in ICS, however, delegate sections or functions before the 5:1 mark. Once this happens, the IC should include an FDISO. If the ICS is handling an incident using groups or divisions, the addition of the fourth group or division is the signal to delegate an FDISO. If the IC delegates the existing groups or divisions into an operations section, this could become the signal to appoint an FDISO. Regardless

Figure 3-10 *Working fires require rapid delegation of safety responsibilities.*

of the type of ICS used, the department should determine the point at which an FDISO is built into the system.

Mutual Aid Request For smaller departments, the appointment of an FDISO whenever mutual aid is called can be a firefighter-saver. Many mutual aid companies work extremely well together (Figure 3-11). But the mutual aid environment asks for firefighters with different cultures, tactics, and equipment to work together. If for no other reason, the FDISO should be appointed to monitor and ensure that the action plan is being carried out by the visiting team as intended. This makes sense for moral reasons, if not for legal reasons.

Firefighter Down/Missing/Injured The seemingly simple incident can get turned upside down and shaken loose when a firefighter emergency takes place. The well-prepared fire and rescue department has a Firefighter Emergency Plan in place to deal with this situation, although they hope never to use it. One key part of this plan should be the immediate appointment of an FDISO to help the IC implement the plan. An FDISO may also be required by the location jurisdiction to start the accident documentation and investigation process.

Incident Commander Discretion Incident management systems recognize that the IC is ultimately responsible for firefighter safety and the incident outcome. At any time

Figure 3-11 *Mutual aid incidents bring together an array of safety concerns.*

and under any circumstance, the IC can delegate the FDISO function. Therefore, a trained FDISO needs to be available for this delegation. Similarly, a department health and safety officer (HSO) or on-duty FDISO should have some discretion to respond to an incident. In these cases, an IC may question or take exception to the FDISO showing up. Hopefully, mutual concern for safety can overcome personality differences and/or turf protection in these cases. A well-written policy can help avoid differences.

Figure 3-12 shows a summary of automatic FDISO response and delegated FDISO response. Appendix B also outlines a sample procedure for FDISO response.

Figure 3-12 *Automatic response versus automatic delegation.*

Incident Safety Officer Utilization

Automatic Response

- Commercial/residential fires
- Wildland/interface fires
- Special-Team incidents
- Target hazards
- Aircraft incidents
- Extreme weather

Automatic Delegation

- "Working" incidents
- Growing span of control
- Mutual-aid incidents
- Firefighter down/missing/trapped

WHERE DOES THE FDISO COME FROM?

So far, this chapter has made a case for the planned response of a trained FDISO. Assume that a fire and rescue department has decided to have preplanned FDISO response. Now a system to get an FDISO on scene is imperative. Who is this person? Where does he or she come from? How is he or she alerted? There are probably as many solutions for these questions as there are fire departments. A sample from around the country shows:

> New York City, New York: Five-person safety battalion. At least one safety battalion member responds to second alarm or greater fires or incidents.

> Springfield, Illinois: FDISO at all incidents (taken from responding officers). Department safety officer (HSO) paged to high-risk incidents.

> Anchorage, Alaska: All company officers may become the FDISO. FDISO appointed as needed by the incident commander.

> California Division of Forestry: Predesignated safety officer part of an overhead team deployed on campaign fires.

> Loveland (Colorado) Fire and Rescue: A team of eight trained fire officers take turns being on call as the FDISO.

This is just a small cross-section of systems designed to get an FDISO on scene. Any one of these or the following options can be tailored to an individual department. The key is to look at each system, look at the department organization, and adapt or alter the system to fit your department and its resources.

Training/Safety Officer(s) On Call

Of all the systems used to get an FDISO on the scene, the use of the Training/Safety Officer (TSO) seems to be the most popular,[2] although it may not be the most effective. In these systems, the TSO can either monitor incident activity and self-dispatch or count on a pager to notify when an FDISO is needed. Some departments have seen the wisdom of splitting these duties. In these cases, usually the Health and Safety Officer is on call to be the FDISO.

This system is most popular because many departments believe that the TSO should be a trained, available fire officer who knows how the department operates. Further, the responsibility of training and safety are in one package—an attractive package to sell in the preparation of budgets and staffing reports. Often, TSOs have radios, vehicles, and other tools that make transition into the FDISO role quick.

The disadvantage to this system is that TSOs are not always available around the clock, putting a burden on the department and the TSO. Often, the TSO is actually the training/safety/infection control/research and development/special projects/recruitment officer for the department. In these cases, the TSO is probably better suited to train a contingent of FDISOs to be available for response.

Health and Safety Committee Members

In this system, the members of the department's health and safety committee are trained to serve as on-duty FDISOs. Typically, an on-duty schedule is made up among the members and the on-duty person monitors the radios and self-dispatches according to department policy. This system is quite popular in small, volunteer departments.

The advantages to this system are that the members are in tune with safety issues and have a forum to communicate; a pool of people to draw from are available; FDISO training is easily accomplished; and an FDISO network is formed.

The disadvantages include: Career (paid) departments may incur overtime due to committee training, meetings, on-call, and actual incident response of members. Extra radios, pagers, and other FDISO equipment must be either purchased for each member or passed from member to member according to the duty schedule.

All Eligible Officers

Strong disciples of incident command systems use this method to delegate an FDISO. In essence, the incident commander appoints an eligible officer to the FDISO position. The eligibility level is established by the department based on which level of officer (lieutenant, captain, battalion chief, etc.) it feels is trained and experienced enough to be an effective FDISO. In order to be proactive, some departments have the FDISO system designed so that the third-arriving company officer or second-arriving battalion chief automatically reports to the IC as the FDISO (if responding as a part of a crew, the officer's crew fills support tasks like running the accountability system, initiating crew rehab, or other like tasks). If this method is chosen, the department *must* provide specific FDISO training to every officer who may fill the FDISO role if the system is to be effective.

The advantage to this system is that the pool to draw from is as large as the department has companies or on-duty battalion chiefs. The officer filling the FDISO role may have a crew to fill critical scene-safety roles like rehab or running the crew accountability system. If an FDISO is not needed by the IC, other assignments can be given or the officer is rotated back into the available resource pool. The disadvantage is that the department must ensure that all officers eligible to fill are trained, which means specialized training for what could be a large group of officers.

Dedicated FDISO

Of all the systems, the dedicated FDISO is perhaps the most desirable. It is difficult, however to commit to this arrangement. The system amounts to a 24-hour duty position where an FDISO is always available. In large metropolitan areas, multiple FDISOs can be on duty to cover the potential for simultaneous significant events.

The advantages are that the FDISO is responding for a specific assignment; the FDISO is positioned to have focused training, experience, and proficiency in the duties of an FDISO; the FDISO can assist with other health and safety assignments for the department. The disadvantage is that this system is hard for most departments to commit to. Those departments with a significant incident injury rate, significant volume of working incidents, or those with high potential for firefighter injury are the departments that typically choose this method.

The key to all these systems is to get a trained, recognized FDISO on scene early.

HOW DOES THE FDISO GET THE JOB DONE?

Define the FDISO Plan

Many of the areas covered in this chapter have offered options to help a fire and rescue department design a system to ensure that an FDISO is utilized at significant incidents. To get the job done, the FDISO needs these options defined by the department and committed in writing for utilization by the line officers. As the popular phrase says: "Don't Just Think It, Ink It!

Where a department has not or chooses not to formalize a system for automated or rapid delegation of the FDISO function, then the FDISO, if ever appointed, will have a difficult task. One area that has not been discussed in designing an FDISO system is that of authority. What kinds of authority should the FDISO have? This subject causes some controversy within the fire service ranks.

The issue is: Should the FDISO have the authority to stop an unsafe act and correct it on the spot? Those who agree that the FDISO should be able to stop an unsafe act are recognizing the fact that the whole purpose of the FDISO is to make the incident more safe. NFPA 1521, Fire Department Safety Officer, 1997 Edition, gives the FDISO this authority to stop, alter, or terminate activities if an imminent threat exists.[3] Those who disagree believe that the FDISO, by stopping or correcting any act on the fireground, is actually countermanding the incident commander. Acts deemed unsafe by the FDISO should be immediately reported to the IC for decision to stop, alter, allow, or correct the situation.

Although Chapter 5 discusses FDISO triggers and traps, one point needs to be made regarding authority—the FDISO can do tremendous damage to an effective program by exploiting the authority given the position. The FDISO in many ways mirrors the work of the incident commander. One can see the battleground here. As with the other components of an effective FDISO system, the authority issue needs to be addressed *prior* to the incident.

Train the FDISO

As with any fire service discipline, a system for initial and ongoing training is essential. Many departments assume that the training/safety officer or any

■ Note
NFPA 1521, Fire Department Safety Officer, 1997 edition, gives the FDISO the authority to stop, alter, or terminate activities if an imminent threat exists.

■ Note
Many departments assume that the training/safety officer or any company officer should know what it takes to fill the FDISO role. This approach is obviously flawed.

company officer should know what it takes to fill the FDISO role. This approach is obviously flawed. Chapter 4 discusses some key training areas to help develop an FDISO. Combine this training with practice in using the FDISO Action Model outlined in Section 2 and the department has an FDISO training program.

Give the FDISO Tools to Do the Job

Several tools can help the FDISO make a difference in getting the job done. The following list is a collection of items that have been used to help FDISOs around the country (Figure 3-13).

Radio As we cover in Section 2, the effective FDISO is constantly roving and watching. A radio is essential to maintaining contact with the IC as well as monitoring the working crews.

High Visibility A fluorescent vest marked SAFETY OFFICER, a green helmet, or other unique identification pays dividends on the incident scene. Not only does it make you easy to spot, but it reminds all crews that safety is important. Many FDISOs report a remarkable change in firefighter behavior just by the presence of the "vest."

Figure 3-13 *Essential incident safety officer tools include proper identification, radio, phone, documentation equipment, and flashlight.*

Personal Protective Equipment (PPE) If the FDISO is to tour the scene, the FDISO may have to cross into hot zones or other areas that require PPE. If the FDISO is assigned to check out interior areas or high-risk areas, the FDISO should partner up and be processed through the crew accountability system. A case can also be made regarding the setting of an example.

Clipboard File Box The FDISO can make a difference if he or she is prepared for those things that the incident commander expects. A metal clipboard file box can give a working surface for notes, sketches or checklists. The file box can contain accident reports, forms, quick reference sheets, and aids such as tablets and pencils. One idea worth considering: Keep a laminated sheet and grease pencil handy so quick signs can be made, such as a LOOK UP sign that can be flashed to crews.

Miscellaneous Cameras, stopwatches, barricade tape, gas detectors or monitors, whistles, and cell phones have been suggested or used by many FDISOs. These and other items may be necessary based on local needs.

Designing an FDISO system can be overwhelming. Hopefully, your department can use this chapter to outline a workable system to ensure that an FDISO is available for significant incidents. Appendixes A and B outline sample FDISO programs that may be useful in assisting you in developing a similar program. Remember, the earlier the FDISO task is filled, the better the chance for a safe operation.

Summary

The design and implementation of an effective incident safety officer function is just as important as having an FDISO on scene. To ensure effectiveness, the FDISO needs to be on scene early in the incident. Therefore, it is best to preplan the FDISO response. Certain incidents require the automatic response of an FDISO whereas other incidents reach a stage where the incident commander needs to delegate the safety responsibility. In both cases, the procedures to get an FDISO in place should be preplanned. Many factors influence the "who" part of assigning the FDISO task. Many advantages and disadvantages appear as a fire department determines who best can fill the FDISO role. Once a system has been developed to ensure that an incident safety officer gets assigned early and often at working incidents, the individuals who will fill this role should be given the training and tools to get the job done and be effective.

Review Questions

1. Explain why the FDISO role should be pre-planned.
2. List four examples where an automatic FDISO response would be beneficial.
3. List four examples where automatic FDISO delegation should take place.
4. List three methods to get an FDISO on scene and discuss the advantages and disadvantages of each.
5. Explain the authorities given to the incident safety officer by NFPA standards.
6. List four tools that can help the FDISO be effective on scene.

Notes

1. A. Washburn, P. LeBlanc, and R. Fahey, "1996 Firefighter Fatalities," *NFPA Journal* 91 no. 4, (July/August 1997), p. 55.
2. This generalization is based on the author's 8 years of traveling the United States in support of incident safety officer training, research, and consulting.
3. NFPA 1521, *Standard for Fire Department Safety Officer* (Quincy, MA: NFPA, 1997).

Chapter

4

Knowing Your Nuts and Bolts

Upon completion of this chapter, you should be able to:

- Discuss the concept of mastery and its benefit to the FDISO.
- List the six dynamic forces that affect fire behavior.
- Define backdraft, flashover, smoke explosion, and free burning.
- List and explain the three factors that influence wildland fire behavior.
- Define blowup and flaring.
- List the three ways to classify fire behavior at an incident.
- List and explain the four general building construction topics that help the FDISO predict building behavior during fires.
- List and explain the seven-step process for analyzing buildings during incidents.
- List the three ergonomic factors that can produce injury.
- List the three strategies to mitigate ergonomic hazards.
- Discuss the two types of thermal stress.
- Explain the role of hydration and nutrition in preventing injuries.

It is one of those shifts when you just know a fire is looming. Eight inches of snow has already fallen, the temperature is hovering around 15°F, and the afternoon commute was stacked with accidents. The early-to-bed firefighters are snoozing and the late-nighters are laughing at the talk show monologue. Off in the watch room you hear the sudden zip of the incident printer while the house lights come on. The station PA clicks, a sharp tone emits, followed by a dispatcher's voice that fills the room; "Attention District 2; Engine 2, Engine 1, Truck 1, Rescue 4, and Battalion 1 respond to a report of fire, Anyplace Nursing Home, 410 E. 5th Street, map page 44J1, Cross of 5th and Jefferson Street. We have an alarm activation and two callers reporting a fire on the third floor. Persons possibly trapped. Time out 2305."

As you move toward the apparatus, your size-up begins with an ugly feeling that this will be a tough one. Every element is stacked against you—an older building, evacuation and rescue challenges, cold weather, fast fire spread, and darkness. You pull up your bunker pants and grin knowing that this is what you have trained for. Then, as an afterthought, you realize you won't have to be the safety officer because you're second-due. You say a silent prayer that the FDISO and incident commander keep everyone safe.

MASTERY

The obvious starting place for any course or text designed to help develop the incident safety officer is an in-depth look at the many hazards, processes, physical dynamics, and operations that can lead to personal injury. For some, this information may seem elementary. The FDISO, however, should strive to achieve mastery of these basics. **Mastery** is the concept that an individual can achieve 90% of an objective 90% of the time.[1] As it relates to the incident safety officer, he or she must have mastered the ability to look at a situation and predict fire behavior, building collapse, and firefighter effectiveness. Additionally, the FDISO needs to master the ability to figure hazardous energy, ergonomics, and human factors into the risk equation.

Each of these topics warrants continual study and application practice if the FDISO is to gain mastery. The FDISO must transition the theory out of the book and apply it on the incident scene.

From an incident commander's perspective, the FDISO is often expected to perform the following and offer judgment regarding:

mastery
the concept that an individual can achieve 90% of an objective 90% of the time

> *Fire Behavior:* Look at smoke conditions and determine the stage of fire, degree of fuel involvement, structural impingement, rate of heat release, and potential for flashover, backdraft, or rapid free burn.

> *Building Construction:* Classify the building as to occupancy type, construction type, and materials used. Predict weak points and collapse potential based on fire location, burn time, imposed loads, and resistance to loads. Establish collapse zones.

Physiology, Kinesiology, and Injury Potential: Predict crew rehab needs, size up ergonomic stressors, and predict the injury potential for firefighters operating in a given environment.

This list can serve as a mini professional development guide for any officer wishing to improve his or her abilities to perform as an incident safety officer. Each of these topics can be researched and studied with many texts. This chapter briefly addresses each of these areas and provides some essential information that can assist in achieving mastery. This chapter does not, however, provide all the information necessary for mastery. Additional texts, curriculums, and formal practice and experience are needed.

■ Note

Mastery comes when the fire officer can look at each component of fire and fire behavior and apply dynamic thinking to actually predict what will happen next with the fire.

FIRE BEHAVIOR

Transitioning from the classroom to reality when discussing fire behavior is a challenge. Typically, the components of fire behavior are presented in a step-by-step curriculum starting with the fire tetrahedron, which is fine for the new and developing fire officer. Mastery comes when the fire officer can look at each component of fire and fire behavior and apply dynamic thinking to actually predict what will happen next with the fire. (Figure 4-1). These components that need to be mastered are:

Figure 4-1 *Mastery of fire behavior helps the FDISO predict dangerous fire spread.*

Fire Tetrahedron

The relationship of temperature, a reducing agent, an oxidizing agent, and the creation of an uninhibited chemical chain reaction make up the fire tetrahedron. Most firefighters know that the removal of any one of the tetrahedron components will lead to extinguishment, but the experienced fire officer knows the degree of removal necessary to extinguish the fire. This is where **fire flow** formulas such as the Iowa State formula or the National Fire Academy adjusted fire flow formula come into play for cooling (see Chapter 9). Likewise, the successful incident safety officer can watch a firefighting effort and tell the effectiveness of the extinguishment effort as it relates to the tetrahedron.

fire flow
amount of water, expressed in gallons per minute, that needs to be applied to a fire in order to absorb the heat released from burning fuels

Fuel States and Classifications

The state or form of a fuel (solid, liquid, gas) as well as its fire classification (A,B,C,D) are considerations in evaluating fire behavior. Perhaps a more pragmatic approach for the FDISO would be to look at the many fuel factors that influence fire and fire spread. Specifically, a given class of fuel, its quantity, relationship to fire exposure, and its burning characteristics are factored into a mental equation in order to predict additional hazard or rapid change of a situation.

Physics and Laws of Nature

Included here are heat transfer (conduction, convection, radiation, and flame contact) as well as laws of heat flow, gas/pressure, and force. Of particular application here are the laws of heat flow and pressure. Ventilation of a hostile interior fire in a building is paramount to successful firefighting. Most rapid increases in fire behavior can be attributed not to the fuels, but to the pressure buildup of fire gases in a container—that is, the room, floor, or entire structure holding the fire. Fuels that are prewarmed accelerate behavior. Recognizing this through a quick evaluation of smoke and gas conditions is essential to predicting behavior.

Properties of Fuels

Properties of fuels include mass and density, flammable ranges and ratios, flash point, ignition point, and boiling point. The chemical aspects of fire behavior are brought to play in this component. The lack of mass in today's construction and manufacturing methods is perhaps the most significant factor in rapid collapse during structural fires. Additionally, understanding flammable ranges can help a fire officer determine if the primary threat to firefighters is a flashover, smoke explosion, or backdraft.

Safety
The lack of mass in today's construction and manufacturing methods is perhaps the most significant factor in rapid collapse during structural fires.

Phases/Stages of Fire Development

Firefighters have developed many labels for the stages of fire: incipient, free burn–steady state, free burn–clear, hot smoldering, cold smoldering, preflashover and postflashover phases. These stages serve as a method to benchmark the fire to prepare for the next phase or stage.

Hostile Fire Events

The understanding, recognition, and rapid prediction of backdraft, flashover, smoke explosion, rollover, mushrooming, blowup, and flaring can make a big difference in the prevention of firefighter injury or death. Likewise, a successful fire fight depends on understanding hostile fire events. Failure to recognize or predict a hostile event can be fatal! Therefore, significant events and their relevant characteristics and conditions are presented here:

Backdraft. A phenomenon that describes the explosive ignition of oxygen-starved, superheated fire gases that are suddenly introduced to air (oxygen). Once backdraft occurs, a resulting free-burning fire ensues, assuming that an oxygen source remains. The size and force of the backdraft is proportional to the size of the "container" that was filled with fire gases. Small, concealed spaces in buildings can create an explosive backdraft that can injure firefighters and collapse a portion of the building.

Flashover. Flashover, the simultaneous ignition of all surface fuels and fire gases within a given compartment, is characterized by a quick-moving flame through the given compartment's space. A room that has experienced flashover may go from an environment with 1,000°F ceiling temperature and 200°F floor temperature to a room with 1,500°F floor to ceiling. Recent investigation and experience is showing that a room flashover may occur with ceiling temperatures as low as 600°F. In these cases, lightweight furnishings, plastics, and other synthetics are thought to be the contributors to rapid flashover.

Smoke Explosion. This term has been argued among fire service personnel. The argument stems from a popular fire service training text that defines backdraft and smoke explosions as synonymous. We know that a backdraft occurs when gases at or above their ignition temperature are introduced to oxygen and an explosive fire ensues. What then, do you call an event where an explosive surge of gases push outward but there is no resulting fire? Many call this event a smoke explosion. The differentiation is necessary because the type and potential for injury changes. For the incident safety officer, a smoke explosion can best be used to describe the sudden expansion and increase in velocity of smoke and fire gases with no resulting ignition of those gases. A smoke explosion in this reference indicates the flash or ignition of some fire gases and the resultant explosive expansion of those gases (these explosions have dropped false ceilings and blown firefighters out doorways). Because the temperature of the gases has not reached their ignition temperature, or because the air–fuel mixture was too rich, a resulting fire does not materialize—quite a difference from a backdraft.

Rollover. Another fire service term, rollover describes the "fingers of flame" that sporadically ignite through a convected column of fire gases and smoke. Rollover is indicative of pockets of fire gases that mix appropriately with air, at the right fuel temperature, to ignite. The flame is then extinguished as the gases slightly cool (because of mixing with cooler gases) or become too rich to burn. Rollover is one of the most reliable signs that a flashover is imminent.

Mushrooming. This is an analogous term used to describe the shape of convected fire gases and smoke as they attempt to rise and escape a compartment. The heated column of smoke and fire gases hit a barrier such as a ceiling then spread laterally. This lateral movement then meets the walls and begins to bank downward, causing a mushroom shape as gases move and eventually fill the compartment. These gases often cool as they become distal to the fire. The cool gases drop and then move back toward the fire's thermal column creating a mushroom-like rotation.

Free Burning. Free burning is the unchecked and sustained advancement of fire across multiple fuels. A free-burning fire continues to increase in size and speed until (1) fuels run out, (2) the available oxygen in the fire compartment is reduced to below 15%, or (3) the fire is controlled by a fire suppression system (sprinkler, dry chem, CO_2), firefighters, or other means. A free-burning fire can become either clear burn or steady state. Steady state refers to a stabilization in the rate of fire spread and heat production, which is the case in continuous fuel beds in the wildland environment or the steady progression of fire through a well-ventilated building. Clear burn is used to describe a fire that is burning fuels "completely." Clear burn is marked by little or no visible color to smoke and fire gases.

Smoldering. Most firefighter references have divided the term *smoldering* into two categories: hot smoldering and cold smoldering. Hot smoldering is fuel and fire gases that are at or above their ignition temperature but cannot ignite due to oxygen deficiency—a candidate for backdraft. Cold smoldering is applied to those situations where open flaming is not possible due to oxygen deficiency and lack of heat. There is, however, enough heat to cause pyrolysis and oxidation of the fuel.

Polyphasic Thinking

polyphasic
multitasking or
multipriority thinking

Seldom are there constants with which to classify a fire, fuel, or its behavior. Likewise, the physical laws that are being applied are never one-dimensional. Applying dynamic and **polyphasic** (multiple task) thinking is essential. As a simple example, to call a working fire in a residential structure a free-burning fire can actually lead to some misunderstanding about potential hazards. The dynamic thinker sees the fire as having compartments that are incipient, some that are free-burning preflashover, some free-burning postflashover. Still other compart-ments are seen as potentially hot-smoldering (especially concealed spaces). Like-wise, the polyphasic thinker sees the effects of heat transfer from multiple media. Conduction, convection, and radiant heat transfer *all* play a role. Various fuels react to the heat and fire phase at different rates.

Behavior Unique to Wildland Fires

Wildfire and interface fires offer some similar as well as unique behavior events. Unlike structural fires, the wildfire environment is heavily influenced by addi-tional factors such as weather and topography. Fuels are even looked at a bit dif-ferently in the wildfire environment.

microbursts
a sudden downburst of winds and precipitation, which can produce straight-line winds of over 100 mph

> *Weather.* Includes temperature, relative humidity, barometric pressure, winds, and weather events such as **microbursts** and tornadic activity.
>
> *Topography.* Factors such as slope (degree), aspect (relationship to the sun), and physical features (chimneys, saddles, barriers, etc.) all influ-ence wildland fire behavior.
>
> *Fuels.* Additionally, wildfire fuels are affected by moisture content, fuel type (ground, aerial, etc.), and continuity of fuels.

These factors affect wildfire behavior with the same dynamic application as discussed in structural fires. Wildland fires experience hostile events just like a flashover in a structure. These events include:

> *Blowup.* Blowup is a wildland fire term used to describe the sudden advancement and increase in fire intensity due to wind, prewarmed fuels, or a topographic feature such as a narrow canyon or "chimney." Sometimes the word *blowup,* so named because it often disrupts or changes control efforts, is used when a ground or surface fire becomes an aerial or crown fire. Because of the sudden increase, a violent convection column causes additional concerns and usually sustains its fire intensity.
>
> *Flaring.* Flaring, a sudden but short-lived rise in fire intensity, can be attributed to wind, fuel, or topographical changes.

CLASSIFYING FIRE

The FDISO needs to evaluate the many foregoing factors and then attempt to classify the fire in terms of firefighter safety. The most practical way to this is to classify the fire behavior as it relates to the potential for change. Seldom is fire behavior a constant. The FDISO can, however, confirm a fire benchmark or anchor event then look forward to the next event that is evident in reading the fire. Then the fire can be classified. Some classifications that are simple and practical from a firefighter safety point of view are:

Stable/Predictable

This classification is good for fires that have shown themselves and are contained within an area with little threat of spread. Likewise, fires that are "surround and drown" may fit this category. A word of caution here: The FDISO who classifies a fire as "stable/predictable" should be on guard and suspicious as this classification may suggest that one can stop worrying about the fire.

Rapidly Changing/Predictable

Any fire that is preflashover should be categorized this way. Other fires that should be classified here include those that are developing significant heat without adequate ventilation and hostile fires in multifuel and wind-fed environments. The essential difference in this category is that the fire behavior environment is changing and the FDISO needs to make quick and successive evaluations of the fire behavior.

Unpredictable

Occasionally, the fire behavior showing is confusing or incongruent to the environment. Further, behavior may take one of several paths based on factors not seen by the fire officer. In these cases, the FDISO should classify the fire as unpredictable. Utmost caution should be employed here. The unstable fire behavior environment requires a high priority for evaluation.

By classifying the fire, the FDISO can prioritize the hazard that behavior presents to the whole firefighting operation. As we know, many other factors (building construction, personnel, etc.) affect the overall safety of the firefighter. These classifications put fire behavior in perspective. Classifying a fire can also be viewed as dangerous just by considering the dynamic nature of fires. The dynamic nature of fire should warn the FDISO to be conservative in the classification of the fire, that is, yield to the worst case classification based on what is visible.

BUILDING CONSTRUCTION

❗Safety
The incident safety officer's greatest contribution to an incident commander is his or her ability to give explicit detail and judgment regarding the collapse of a given building being attacked by a given fire.

Without a doubt, the incident safety officer's greatest contribution to an incident commander is his or her ability to give explicit detail and judgment regarding the collapse potential of a given building being attacked by a given fire. In order to make this judgment, the FDISO must draw on significant knowledge. The fire officer that draws from "experience" to predict building collapse is fooling himself into a situation where a building will collapse without warning. Francis Brannigan, noted author of *Building Construction for the Fire Service* says it best when he says: "Relying on experience alone is not sufficient. Firefighters must be aware of the theories and principles involved (in building construction)."[2] To say that no building collapses without warning is erroneous. The warning for structural collapse is the

Figure 4-2   *The building construction student applies his or her knowledge to better predict collapse.*

FDISO's ability to understand building construction and the effects of fire on the building. Granted, once collapse occurs, unpredictability enters the picture.

Those who have studied the Brannigan text are better prepared to make judgments regarding collapse potential. This FDISO text is not a substitute for an in-depth study of the building construction books that detail the subject. This book does, however, take some essential building construction elements and thought processes and show how to evaluate and offer judgment on collapse potential (Figure 4-2).

Key Topics

Predicting building collapse is dependent on the application of some essential building construction topics as listed in Figure 4-3 and explained briefly herein.

Imposition and resistance of loads. The basis of all building construction techniques is to carry a load to earth. The load itself must be described or classified. The direction or application of a load to a component is called the imposition of the load. The material that has a load imposed upon it must resist the load. Figure 4-4 shows examples of each of these.

Figure 4-3 *To achieve mastery of building construction and to predict collapse these topics must be studied.*

Key Building Construction Topics

- Imposition and resistance of loads
- Characteristics of building materials
- Building assembly components
- Construction classifications
- Effects of fire on buildings

Characteristics of building materials. The materials used to resists loads have unique characteristics. The fire service looks at how these materials react during a fire and how a material's ability to resist a load changes during fire conditions. Figure 4-5 looks at some common characteristics of building materials.

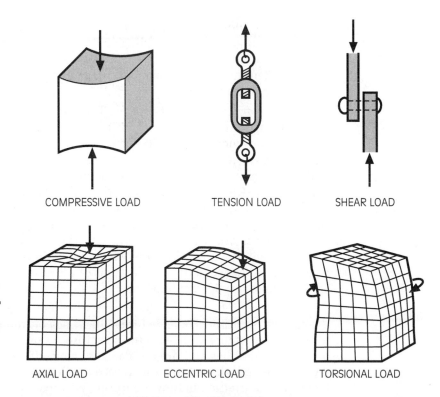

COMPRESSIVE LOAD   TENSION LOAD   SHEAR LOAD

Figure 4-4
Understanding how loads are imposed on and resisted by materials is essential to predicting collapse.

AXIAL LOAD   ECCENTRIC LOAD   TORSIONAL LOAD

IMPOSITION AND RESISTANCE TO LOADS

Material	Load Resistance	Effects of Fire or Material
Wood	Marginal resistance to compression, tension, and shear.	Burns and loses strength.
Steel	Excellent resistance to compression, tension, and shear.	Noncombustible. Elongates when heated (1,000°F). If restrained, steel will twist in order to elongate when heated. Cold drawn steel such as cable and pins fail at 800°F.
Concrete	Excellent resistance to compression. Poor resistance to tension and shear unless reinforced with steel.	Noncombustible. Can absorb great quantities of heat prior to strength loss. Reinforced concrete may fail early if steel is exposed to heat.

Figure 4-5

Characteristics of common building materials.

Building assembly components including truss emphasis. The components used to assemble a building are numerous. Some basic elements, however, are used by all methods. Columns and various types of beams are the essential core of every building. Figure 4-6 lists basic building component terminology.

General construction classifications. The general classification of a construction method helps the fire officer better understand the materials, methods, and components present in a given structure. There are many different ways to classify the construction of a building. Codes define building construction one way, contractors another. Even the fire service classifies construction styles differently within the profession. The important point regarding construction classification is not in the method that you choose, it is in your ability to look at a building and understand how the building is assembled and what materials are being used. Then you should be able to apply your knowledge of material characteristics, load resistance, and special or weak features of the construction style that will enable you to predict a collapse.

Effects of fire on buildings. Fire essentially attacks the materials used to assemble a building. The affect this attack has on the building is related to the fire-resistance built into the structure. In most cases, fire degrades building assemblies and fire resistance systems. In addition to fire degradation, fire growth and fire and smoke containment must be considered key topics in predicting how a building will react to a fire—understanding these topics help the FDISO predict collapse.

Column	A structural member that is loaded in compression.
Beam	A structural member that transmits a load perpendicular to the load. This places the top of the beam in compression and the bottom of the beam in tension. Beams may be fixed on one end, both ends, or any other combination.
Girder	A beam that supports other beams.
Lintel	A beam that supports the load above an opening in a wall.
Truss	Essentially a beam that uses triangled struts and ties to attach a top and bottom chord, simulating a solid beam.
Wall	Essentially a long, narrow column.

Figure 4-6 *Common building components.*

Analyzing Buildings During Incidents

With a keen understanding of the foregoing topics, the incident safety officer can analyze buildings during structure fires. This analysis is necessary to establish if the building will "behave" during fire fighting operations.

Listed here is an easy, seven-step process that will help the FDISO analyze buildings (and other structures) and predict collapse potential (Figure 4-7). This process can be equally applied to other structures such as bridges, cranes, mine shafts, and wells. It is important to note here that any building analysis during an incident should be cyclic, that is, performed on a regular basis as conditions and time change.

Steps

1. Classify the type of construction (Figure 4-8). The quickest way to classify the type of construction is to use standard language such as wood frame,

---

Seven-Step Process to Analyze and Predict Building Collapse

1. Classify the type of construction.

2. Determine the degree of fire involvement.

3. Visualize load imposition and load resistance.

4. Evaluate time as a factor.

5. Determine the weak link.

6. Predict the collapse sequence.

7. Proclaim collapse zones.

Figure 4-7 *These seven steps can save needless firefighter fatalities!*

Figure 4-8 *A building may be classified by more than one construction method.*

ordinary, steel, and so forth to provide a quick reference for all responders and to underscore the characteristics of that construction type.

2. Determine the degree of structural fire involvement (Figure 4-9). The determination of whether a fire is a "contents" or "structural" fire is imperative. Too often, fire departments use the term "structure fire" to dispatch crews. True, this term may serve as a warning, but many fires are merely contents fires. Many fire service leaders have suggested that crews should be dispatched to a "fire in a building." Only upon arrival and size-up is a fire classified as structural. For the sake of this book, "structural fire" means that the load-bearing components of a building are burning.

Figure 4-9 *Once load-bearing structural members are attacked by fire, collapse may come quickly.*

Figure 4-10 *The FDISO should mentally "undress" the building and visualize load resistance.*

The FDISO must look for signs that structural members are being attacked by fire. Fire in concealed spaces, contents fires in unfinished basements, attic fires, or heated exposed beams and trusses, are all examples of fires that have become structural. The point here is obvious: Once the structure is involved, the attention to collapse should be immediate.

3. Visualize load imposition and load resistance (Figure 4-10). This step is more an art than a science. To practice this art, the FDISO visually scans the building, tracing any given load to the ground. In doing so, the FDISO determines if there is any structural element that is carrying something it should not and whether key elements are being attacked by fire. This step helps the FDISO catch some high-hazard structural components or conditions that have typically factored in firefighter deaths after a collapse. Examples of this include HVAC units, signs, heavy stock, overhangs, open-spans, and unsupported loads.

4. Evaluate time as a factor (Figure 4-11). Some structural elements fail as soon as fire (heat) has reached the material. Other materials absorb incredible heat for a long duration before they become susceptible to collapse. Time as a factor should be brought into the collapse equation. Often, the FDISO has no idea how long a fire has attacked structural components. In these cases, it is best to err on the side of rapid evacuation.

Although no formula exists that can predict collapse time, a few truisms can be applied regarding time. These are:

- The lighter the structure, the faster it will come down.
- The heavier the imposed load, the faster it will come down.
- Wet steel buys time.
- Gravity and time are constant, resistance is not.

5. Determine the weak link (Figure 4-12). Often, the collapse of a building is the result of one specific structural failure—the weak link. Each type of building construction has its own unique weak link, which is why study of building construction texts is so valuable.

Most weak links are small, lightweight components that have little protection from heat. A few typical weak links include:

Trusses (floors/roofs). Trusses are nothing more than fake beams. Extremely light materials, a geometric shape (triangles), and open space are used to replace a solid beam. This combination is deadly when fire is introduced. Other questions arise when dealing with trusses: Does the truss span a significant distance? What materials are being used to piece together the truss? NOTE: Gusset plates typically hold together trusses. Gusset plates pop out quickly during rapid heating and fall off completely when a small portion of the truss woods burns away in as little as 5 minutes of burn time. Each of these questions can lead to the point of weakness that will precipitate a collapse.

Connections. Structural failure is often the result of the failure of a connection. Connections are found where different components are brought together, as when a beam attaches to a column. Usually, the connection (bolt, pin, weld, screw, gusset plates, etc.) is made of a material with lower mass or fire resistance than the assembly it connects. In some cases, the connection relies on gravity and an axial load to hold it in place. A shift or lateral load may cause the connection to fail. If a connection of two steel beams and a steel column is rigid (bolted or welded) and the column is restrained, a heated beam will begin to twist in an attempt to elongate, thus dropping whatever weight was on the beam.

Safety
Gusset plates typically hold together trusses. Gusset plates pop out quickly during rapid heating and fall off completely when a small portion of the truss woods burn away, in as little as 5 minutes of burn time.

Figure 4-11 *Once fire has entered the attic, time keeping becomes paramount.*

Figure 4-12 *The emphasis on trusses as a "weak link" can not be overstated.*

Overloading. Many buildings are used for occupancies for which they were never intended. An example is a doctor's office in a remodeled home. Walls are removed, increasing floor and roof spans. The massive weight of medical files, office equipment, lab, and maybe even X-ray equipment is added. Remodeled buildings should always hold a concern of overloading. Attic spaces are seldom designed to be used as storage areas but the open space between trusses often lures the occupant into storing old files, parts, or furniture. Two feet of wet snow may not over-stress a flat roof in Durango, Colorado, but that same snow in Amarillo, Texas, could collapse many roofs.

6. Predict collapse sequence (Figure 4-13). Once the above steps have been completed, the FDISO can visualize a collapse scenario for the building involved. Typically, a building experiences a partial or general collapse. A partial collapse is one in which the building can accept the failure of a single component and still retain some margin of strength. Often, a partial collapse is partial because another component picked up the weight of what fell (like interior partition walls).

In a general collapse the building fails to absorb the failure of an element and succumbs to gravity. Predicting collapse also includes a determination if walls will fall inward or out. Likewise, an inward fall of a wall may cause the bottom of the wall to kick out. The failure of a roof assembly may weight interior partition walls that were not designed for the load. The slightest movement could cause an explosive release and further collapse.

7. Proclaim collapse zones (Figure 4-14). This step is nothing more than communicating the building analysis process. It is here that firefighter lives can be saved. Likewise, the incident commander can make tactical changes based on the collapse zones established through this process. In so doing, it will be

!Safety
The failure of a roof assembly may weight interior partition walls that were not designed for the load. The slightest movement could cause an explosive release and collapse.

Figure 4-13 *The FDISO should visualize the potential collapse prior to an actual collapse.*

difficult to claim that a building "collapsed without warning." Not only have crews been warned, but they have been made aware of where the building will collapse.

PHYSIOLOGY, KINESIOLOGY, AND INJURY POTENTIAL

There are few peacetime occupations that stress the human body to the degree of that experienced by firefighters in a hostile, working fire environment. Fire suppression imposes stress on the physiological and kinesiological components of the firefighter. *Physiological* components include the metabolic processing (fueling) of the firefighter. *Kinesiological* components include the muscular motion and activity of the firefighter. Further, the problems associated with human kinetics in an environment is called *ergonomics*. The firefighter who has not addressed each of these areas prior to an incident will be injury prone because of the arduous nature of fire suppression. The Incident Safety Officer must look at both the firefighter and the environment to determine if physiological or ergonomic considerations must be addressed to reduce the potential of an injury. This task is not easy. The firefighters on scene can range from a well-conditioned athlete to a heart-attack-waiting-to-happen.

Enough cannot be said about the need for firefighters to be strong, flexible, and aerobically fit. Further, the firefighter should adopt a program of efficient fueling for his or her metabolism. Much has been written about firefighter fitness and nutrition that can help achieve these aims. Unfortunately, the FDISO is not in the position to effect physical fitness improvements while at the incident scene. The FDISO can, however, make a difference in the way he addresses ergonomics and human performance. A solid foundation in these areas can help to quickly identify factors that add to injury potential and enable the FDISO to quickly prevent or address each of these factors.

■ Note
Unfortunately the FDISO is not in the position to effect physical fitness improvements while at the incident scene. The FDISO can, however, make a difference in the way he addresses ergonomics and human performance.

Figure 4-14 *Collapse zones should be marked and communicated to working crews.*

Ergonomics

ergonomics
science that seeks to adapt work conditions to that of the worker in order to prevent injuries or make the environment safer

The study of **ergonomics** has only recently taken precedence in the workplace. For firefighters, the workplace includes the fire station as well as an incredible variety of environs. To have an engineered workplace that is "ergonomically friendly" is ludicrous at best for the firefighter working an emergency scene (Figure 4-15).

Being aware of the stresses and strains that face firefighters is the key to preventing injuries. Often, the FDISO needs to say "Time out" and adjust the way our people have chosen to tackle the situation. To do this, the FDISO should evaluate the environment, the relationship of the worker to the environment, and the task being attempted. Often, a slight change to any of these three areas can reduce injury potential. Each of these areas can be evaluated to determine if an injury potential exists.

■ **Note**
The FDISO should evaluate the environment, the relationship of the worker to the environment, and the task being attempted. Often a slight change to any of these three areas can reduce injury potential.

Physical environment. Included in this area should be an examination of the surface conditions of the working area. Obviously, secure footing is essential to safe operations. Temperature variations, close proximity to equipment, and lighting integrity also should be considered. Often, distractions such as noise, flashing lights, swinging mechanical equipment, and weather extremes are part of the work environment.

Relationship of the worker to the environment. In a given environment, the worker might be forced to tackle a task in a bent-down or stooping position. Other times the worker must ascend or descend to accomplish a task. Pulling, twisting, and pushing are other actions that may cause

Figure 4-15
*Firefighters
seldom find an
"ergonomically
friendly"
workplace.*

injury. The speed or pace at which the task is completed may also create
an injury.

Task. The energy and the amount of focus or attention required to complete a task should be considered. Sometimes, the task priority is such
that additional stress is created. In other cases, the number of people
available to accomplish the task is the primary factor leading to injury.
The types of tools and equipment necessary to accomplish the task also
should be considered.

Once the physical environment, relationship of the worker, and the task
have been evaluated, certain hazards will be obvious. Then the FDISO can utilize
one of three strategies—awareness, accommodation, and acclimation—to abate or
mitigate the hazard.

Awareness. This strategy acknowledges that the worker is less apt to suf-
fer an injury if he or she is keenly aware of the problem, thus exercising
greater caution. This form of injury reduction is the most simple.

Examples include the warning of a slippery surface or a reminder to lift with one's legs.

Accommodation. Injury potential can be reduced by accommodation, that is, by altering the environment or the task. The use of protective clothing or a lumbar support belt are forms of accommodation. Placing a roof ladder on a pitched roof is a method of accommodating the worker in an environment that requires ascending and descending on a diagonal surface. Adding more people to a task is a form of accommodation. Artificial lighting helps accommodate darkness.

Acclimation. This strategy is the most difficult to implement during an incident because most acclimation is done proactively. Physical fitness programs that include strength and flexibility training are examples of acclimation. One form of on-scene acclimation would include the rapid rotation and rehabilitation of crews. Having firefighters drink water and perform simple stretching exercises prior to an assignment are other examples of acclimation.

Human Performance

The study of human performance, and more recently, firefighter performance, is well documented in text, trade magazines, and medical journals. One conclusion is reached by all: Fire fighting requires physically fit humans. Those who are not fit for duty present an injury risk to themselves as well as others due to the team nature of fire fighting. Conversely, the physically fit firefighter can work longer, harder, and faster with minimal injury risk. Both the fit and unfit firefighter can become fatigued. Regardless of the level of fitness, the fatigued firefighter creates an equal injury risk. The FDISO can help prevent injuries while on scene if he or she applies knowledge of human performance factors to the incident scene in order to prevent fatigue, and therefore, injury. These factors include thermal stress, dehydration, and energy depletion.

Thermal Stress Thermal stress can come in the form of heat stress and cold stress. Heat stress can be further divided into internal or metabolic heat and external or environmental heat. The human body is actually quite amazing in that it has many systems to help regulate a normal core temperature of 98.6°F. Once the body temperature rises above this number, heat stress is created. Factors that may cause core temperature to rise include:

- Activity
- Humidity
- Air temperature
- Effectiveness of cooling mechanisms
- Sun, shade, and wind

Full protective clothing reduces the body's ability to evaporate heat thru sweating. Likewise, high humidity reduces the evaporation of sweat. Working in direct sunlight adds additional heat stress, especially at higher altitudes. Firefighters suffering from heat stress go through a series of heat-related injuries that get progressively worse as heat builds. Figure 4-16 outlines these heat injuries.

Cold stress is similar to heat stress in that a series of injuries can occur if the body's core temperature cannot be maintained. In this case, however, the body temperature is being reduced or cooled. Moisture becomes an enemy in cold stress. The perspiration that a firefighter experiences while fighting an interior fire can cause shivering and lack of concentration when the firefighter moves outside to freezing weather. Other factors that affect cold stress include air speed and temperature (wind chill), level of activity, and the duration and degree of exposure. Figure 4-17 outlines cold stress injuries.

Thermal stress injuries can be prevented through accommodation, rotation, and hydration. Accommodation includes the use of extrawarm clothing during cold extremes or the use of misting fans during extreme heat. Rotation is the aggressive action of rotating crews through rest, heavy tasks, and light tasks in order to minimize the stress caused by working in extreme environments. Hydration cannot be overemphasized in heat stress environments, but it is also effective, though often forgotten, in cold stress environments. The importance of hydration is discussed in the following section.

Safety
○ Thermal stress injuries can be prevented through accommodation, rotation, and hydration.

Dehydration Water is vital to the peak operation of virtually every body system from transport of nutrients, to blood flow, to waste removal, to temperature regulation. When the body becomes dehydrated, these systems start to shut down in order to protect themselves. With this shutdown comes fatigue, reduced mental ability, and, in extreme cases, medical emergencies such as renal (kidney) failure, shock, and death. The working firefighter must accommodate for the wearing of heavy

Heat Stress Injuries	
• Heat stroke	Medical emergency! Marked by hot, flushed, and *dry* skin.
• Heat exhaustion	Serious injury. Person is dehydrated, skin is cold and clammy. Person may be weak, dizzy, and nauseous.
• Heat cramping	Painful muscle spasms. Typically legs and arms are affected.
• Heat rash	May be an early warning sign.
• Transient heat fatigue	An early sign characterized by physical exhaustion. Remedied by rest and hydration.

Figure 4-16 *The FDISO should be observant of signs and symptoms of heat stress.*

Cold Stress Injuries	
• Hypothermia	Can range from mild to severe. Mild cases are marked by shivering and loss of coordination. Lethargy and coma can onset quickly.
• Frostbite	A serious "local" injury meaning that a body part is frozen.
• Frostnip	A local injury. Most people do not realize they have frostnip. It is, however, a precursor to frostbite.

Figure 4-17
Hypothermia is the cooling of the body's core temperature. This condition should be avoided at all cost.

!**Safety**
Firefighters should drink a quart of water an hour during work periods, optimally in 8-ounce portions spread over the hour.

clothing that does not allow evaporation of sweat. Additionally, the firefighter must account for strenuous physical activity under stressful situations. Hydration of firefighters should be paramount, even to the point of excess.

As a rule, firefighters should strive to drink a quart of water an hour during work periods, optimally in 8-ounce portions spread over the hour.[3] Substituting soda or other liquids for water can slow the absorption of water into the system. For this reason, just water should be given for the first hour. For activities lasting longer than an hour, some consideration can be given to adding essential electrolytes and nutrients along with the water. Many sports drinks are available for this purpose. These sport drinks are best diluted 50% with water in order to speed their absorption into the system.

Energy Depletion Although dehydration and thermal stress can lead to energy depletion, most firefighters associate energy depletion with the need for food. Nutrition or fueling of the firefighter can rejuvenate the firefighter or put the firefighter to sleep. Too often, rehab feeding efforts accomplish the latter. The FDISO should view firefighter fueling from the perspective that a firefighter properly nourished will work smarter and safer. The firefighter who is improperly fed will not only want to "crash," but will likely make sluggish mental calculations leading to injury.

So what is the proper way to nourish firefighters? A study of essential nourishment theory can help answer this question. In basic terms, maximizing energy from the human machine takes a balance of four essential elements: oxygen, water, blood sugar (from food), and insulin. When this balance is present, other essential hormones and enzymes are created that makes for a well-running human machine. The element that we often misprescribe is food—the foundation for building balance. True, *all* food is fuel for the firefighter. That is, all food is converted to glucose (blood sugar) for the muscles to use as fuel. Insulin is released into the bloodstream to help convert the blood sugar into energy for muscle use. Some foods are digested faster than others though. Complex carbohydrates such as breads, pasta, and beans require tremendous amounts of blood, and time, for digestion. This fact explains why some firefighters want to go to sleep after eating:

Blood is needed for digestion and less energy is available for thinking and physical tasks.

The key to providing quick energy to the firefighter is to find a balance of protein, fat, and carbohydrate. Ideally, this balance should be a 30/30/40 mix—30% protein, 30% fat, and 40% carbohydrate.[4] This balance provides essential elements from three food groups. A balanced approach achieves a few benefits. One, the balance stabilizes insulin release into the bloodstream, helping to reduce the roller-coaster of blood sugar levels that often leads to sporadic activity, chemical imbalance, and fatigue. Second, the balance approach stimulates the release of hormones and enzymes that optimize physical and mental performance.

Choosing the best protein, carbohydrate, and fat also promotes steady, sustained performance. Protein is best derived from low-fat meats such as turkey, chicken, and fish. Eggs and cheese also offer protein. Fats should be of the monounsaturated type as found in olive oil, nuts, and peanut butter. Often, carbohydrates are dangerous in that so many of the foods we typically find on the fire scene are rich with unfavorable carbohydrates. Candy, bread, potatoes, and bananas are all carbohydrates that are quite rich and have a tendency to slow down the worker. Good carbohydrates include green vegetables, apples, tomatoes, oranges, and oatmeal.

The department that preplans nourishment for rehab efforts can achieve the balance outlined above. Menu planning with a nutrition or fitness expert is one way to preplan rehab. In cases where a department has not preplanned rehab nourishment, the FDISO can make recommendations based on the balance approach. A loose sampling of nourishment for rehab might include:

> **Morning:** Breakfast burritos (scrambled egg whites, chopped Canadian bacon, low-fat cheese, flour tortilla), water, and apples. One Colorado fire department has these premade and frozen. Upon request, the burritos are microwave-thawed and heated, then brought to the scene.

> **Afternoon and evening:** Turkey and cheese sandwich (use thinly sliced bread to reduce heavy carbohydrates), water, apples, and small portions of peanuts or cashews. It is better to have high-piled turkey than to eat two or three sandwiches because the extra bread can easily throw the balance of this suggestion.

> **Anytime:** Commercially available energy bars and water. It is important to make sure that the energy bars are the 30-30-40 balanced type.

The rehab effort that includes balanced nutrition as previously stated, as well as substantive hydration, helps keep firefighters performing well both mentally and physically, and helps reduce injury potential.

Summary

In order to be effective during incidents, the FDISO must have a solid foundation of technical understanding of fire behavior, building construction, and human performance. Fire behavior is key because of the potential of a fire to spread rapidly and cause serious injury to firefighters. By applying dynamic thinking and classifying the behavior characteristics of a fire, the FDISO can better prevent the serious injuries. One of the first questions that incident commanders ask the FDISO is when a building is going to collapse. The FDISO who can predict collapse will prevent deaths. By applying a simple building analysis and a thorough understanding of key building construction topics, the FDISO can make this prediction. Further, the FDISO needs to understand the many stresses the human body endures as a firefighter. Ergonomics, thermal stress, dehydration, and energy depletion can all add to injury. Many times, the FDISO can influence a slight adjustment in the environment to prevent injuries. Maximizing rehabilitation efforts should be a thrust of the FDISO.

Review Questions

1. Define the concept of mastery. Discuss the importance of mastery to the FDISO.

2. List the six dynamic forces that affect fire behavior.

3. Define backdraft, flashover, smoke explosion, and free burning.

4. List and explain the three factors that influence wildland fire behavior.

5. Define blowup and flaring.

6. List the three ways to classify fire behavior at an incident.

7. List and explain the four general building construction topics that help the FDISO predict building behavior during fires.

8. List and explain the seven-step process for analyzing buildings during incidents.

9. What are the three ergonomic factors that can produce injury?

10. What are the three strategies to mitigate ergonomic hazards?

11. List the two types of thermal stress and explain how to prevent both.

12. Explain the role of hydration and nutrition in preventing injuries.

Notes

1. I.K. Davies, *Instructional Technique* (New York: McGraw-Hill, 1981), p. 22.

2. F. Brannigan, *Building Construction for the Fire Service* 3rd ed. (Quincy, MA: National Fire Protection Association, 1992), p. 12.

3. U.S. Fire Administration, *Emergency Incident Rehabilitation,* FA-114 (Washington, DC: USFA Publications, July 1992).

4. B. Sears, *Enter the Zone* (New York: HarperCollins, 1995), p. 71.

Chapter

5

Triggers, Traps, and Incident Commander Relations

Learning Objectives

Upon completion of this chapter, you should be able to:

- List four methods to help trigger incident safety. Explain the appropriateness of each.
- List the three FDISO traps and discuss how each can render the FDISO ineffective.
- Describe the four components that support positive incident commander relations.

The fire has only consumed two homes and is still burning freely on a hill-side full of light grasses and scrub oak. The weather will hold and winds will be stable for the next 3 hours. The current action plan, which is to continue a pincer attack with brush patrol apparatus and progressive hoselays from engines, appears to be appropriate. Two houses are in the path but it looks like we can knock the fire if we stay aggressive. One look at the firefighters in rehab tells a different story though. They are being spent quickly. As you approach the incident commander to report your update, you frame in your mind a way to suggest the need for relief crews if we are to remain effective.

"Chief, the weather report I received is favorable for what you've got going here—good for the next 3 hours at least. I'm not comfortable though with the number of firefighters present. They looked whooped."

"I think we got it. We don't need any other crews. Get over to rehab and tell those guys to pick it up!" orders the incident commander as he turns away from you and starts talking in his cell phone. You stare at the back of his day-glo orange vest and ask yourself, "Now what? Maybe there is a different way I could approach this."

Thus far, a case has been made for a technically competent fire department incident safety officer (FDISO). However, technical competence alone does not create an FDISO who makes a difference. The element that tips the scale toward making a difference is the ability to communicate clearly and appeal to the safety sense that is often hidden during working incidents.

By and large, firefighters are a proud, strong, and reasonable group who thrive on competition, adrenaline, and challenge. The FDISO can trigger favorable or unfavorable responses when confronted with a group of firefighters driven by challenge. Often, the new or inexperienced FDISO falls into some common traps that may seem trivial, but that thwart the goal of making the incident safer. Further, if these traps are not addressed, the long-term effectiveness of an individual serving as an FDISO will be minimal. Likewise, the FDISO must package concerns in a way that is appropriate and appealing for the incident commander (IC). Failure to work in harmony with the IC is also a failure in firefighter safety. This chapter looks at ways to trigger favorable results, avoid ineffective traps, and improve incident commander relations.

> **■ Note**
> The FDISO can trigger favorable or unfavorable responses when confronted with a group of firefighters driven by challenge.

> **■ Note**
> Failure to work in harmony with the IC is also a failure in firefighter safety.

TRIGGERS

The incident scene is much like a championship football game in that everyone present is trying to accomplish a goal. To accomplish the win, a coach must develop a strategy for victory and assign tactics (plays) to meet that strategy. The coach must harness the varying skill, talent, fortitude, and emotional level of his players to achieve the win. Although everyone wants to win, occasionally a player operates outside the coach's plan, resulting in a loss.

When fighting fire, the incident commander utilizes crews that possess a variety of skill, knowledge, talent, and emotions. Occasionally, a player (firefighter) operates outside the plan. This firefighter wants to "win," however, the emotion, skill, talent, or method is not appropriate for the incident commander's plan. Unfortunately, the loss here is not just losing a game; the loss is likely to include the loss of life or the loss of a working firefighter to injury. Here the FDISO can step in and make a difference. By using certain triggers, the FDISO can remind individuals that we, as a profession, operate according to a plan, in a safe manner, within solid risk/benefit categories. "Work safe" triggers fall into a hierarchy of effectiveness based on the situation at hand.

Visibility

The incident safety officer should wear a high-visibility vest that clearly states "SAFETY" (Figure 5-1). Although this trigger may seem simple, it is surprisingly effective. In many cases, just seeing the word "SAFETY" is enough to imprint safe ethics and habits. In many ways, this is like the power of suggestion employed by many advertisers. Often, firefighting crews change an unsafe procedure or condition upon seeing the safety officer. This self-correction is desirable on scene in that the FDISO can concentrate on other items of concern.

Figure 5-1 *Being visual to everyone by the use of a SAFETY vest can help trigger safe behavior. (Photo courtesy of Richard W. Davis.)*

Figure 5-2 *On the rare occasion that an FDISO must go inside an IDLH environment, the FDISO should partner up and use full PPE.*

Example

Philosopher, theologian, and Nobel Peace Prize winner Albert Schweitzer once said, "Example is not the main thing in influencing others. It is the only thing."[1] The effective FDISO uses the example trigger to instill safe behavior in others. Often, the small habits that an incident safety officer performs influence others. To illustrate, the FDISO should always, without fail, participate in the crew accountability system. Likewise, the FDISO should use appropriate personal protective equipment (PPE), follow department policies, obey hot zone/cold zone markers, and utilize a partner and backup for immediately dangerous to life and health (IDLH) environments (Figure 5-2).

Soft Intervention

Humor, subtle reminders, information sharing, and peer talk are examples of soft interventions (Figure 5-3). These are employed as a way to trigger safe behavior when the witnessed action is not necessarily life threatening. Even in some significant safety concern environments, this trigger can be effective if the person getting the message is a peer, a senior member, or someone who typically operates in a safe, professional manner. In most cases, this method of delivery accomplishes the change you wish, especially if the FDISO acknowledges the wisdom or choice made previously then interjects a third interpretation or additional information to underscore the safety concern.

Interjecting humor is effective when making soft interventions, but caution is warranted. The humor must not trivialize the safety concern. Humor used to

Figure 5-3 *Often, a simple reminder (soft intervention) is all that is needed to prevent an injury.*

make a safety point is best when the environment allows face-to-face communication and centers on the circumstance and not the action of the firefighter.

Firm Intervention

■ Note
The firm intervention is more or less an official order to stop, alter, or change an action.

■ Note
Anytime a firm intervention is used, the FDISO should immediately relay the concern to the incident commander.

Firm interventions are best used when fast action is required to eliminate a life-endangering situation (Figure 5-4). In less serious situations, a firm intervention may be a necessary trigger for an individual who repeatedly creates or is involved in unsafe situations. Likewise, a firm intervention may be necessary when a soft intervention is ineffective in accomplishing the suggested change. The firm intervention is more or less an official order to stop, alter, or change an action. This intervention could lead to power issues on the incident scene. One way to avoid conflict is to make the intervention with clear and direct language and by taking complete responsibility. For example, an FDISO who witnesses a crew operating in an imminent collapse zone could relay via radio, "Attack Team 3, Safety. Evacuate that position immediately. That is an imminent collapse zone. Acknowledge." Anytime a firm intervention is used, the FDISO should immediately relay the concern to the incident commander.

Figure 5-4 *Firm intervention is designed to prevent an immediate life-threatening hazard. (Photo courtesy of Richard W. Davis.)*

Matching the intervention to the degree of concern is essential to achieving buy-in with the person or crew in question. If the intervention is viewed as irrelevant or demeaning to the crew, change may not occur.

TRAPS

The fire department incident safety officer can find him- or herself getting trapped into operational modes and activities that will render the FDISO and any incident safety officer program ineffective. Often, the new or inexperienced FDISO wanders into these traps. Obviously, the FDISO is wise to avoid these traps.

The Bunker Cop

The bunker cop syndrome is marked by the FDISO who spends much of his or her time looking for missing, damaged, or inappropriate use of personal protective equipment (PPE) (Figure 5-5). Typical is the FDISO who is always asking working firefighters, "Where are your gloves?" or "Why isn't your helmet on?" Like a traffic cop, the bunker cop is focused on one specific component of safety—protective equipment. Unfortunately, this focus causes the FDISO to miss the big picture of incident safety. Further, this approach will eventually alienate the FDISO and the FDISO program because issues of protective equipment are most suitable for the company officer, group supervisor, or team leader level. If these leaders take care of PPE issues, the FDISO is no longer needed to patrol PPE issues.

Figure 5-5 *The FDISO who takes a "bunker cop" approach will miss the big picture.*

Granted, the FDISO who fails to recognize situations where firefighters are not properly protected is also negligent. In these cases, the FDISO should report the infraction to the person's team supervisor as soft intervention. If the PPE issue keeps resurfacing at incidents, the FDISO may be faced with a situation where department values need to be shifted toward more routine use of PPE. In these cases, the FDISO needs to be a bunker cop short term while working on the departmentwide issue of moving the bunker cop responsibility to the company level.

Another form of bunker cop is the FDISO who focuses on skill proficiency. It is quite easy to look at a crew throwing ground ladders and determine if they are doing it by the book. The FDISO who brings up skill deficiencies may not be received well by working crews. One incident comes to mind where an FDISO made this mistake. The senior officer who was operating as a company officer found the FDISO intervention very annoying. He promptly put the ladder down and said, "Here, you do it smarty pants!" One way to avoid this predicament is to ask yourself, what is the likelihood of an injury if they continue to do it that way, and how serious will that injury be (remember risk management from Chapter 2?).

New or inexperienced FDISOs often fall into this trap because of the familiarity and comfort of basic PPE and skill evolutions. When the new FDISO was (or acts as) a company officer, PPE and crew safety is automatic, therefore, a comfort zone is established and these comforts come out in the new FDISO. Although it may be uncomfortable, the new FDISO needs to jump past PPE issues and look at big picture items such as fire behavior, building construction, and risk management.

The CYA Mode

The well-known mnemonic CYA best describes the incident safety officer who spends an inordinate amount of time ensuring that he or she will not personally be held accountable for incident scene actions. Some may find this label insulting; reality, however, must be acknowledged. The proliferation of NFPA standards, OSHA Code of Federal Regulations (CFRs), and local requirements have placed the FDISO in a position of liability in regard to following due diligence outlined in these codes and standards. It is easy to imagine an incident safety officer being found liable following a significant firefighter injury or duty death. It can easily be implied that no death should occur if an incident safety officer is on scene. This proposed scenario can lead to a CYA approach by the FDISO.

A CYA approach is witnessed in the FDISO who constantly uses CFR, standards, and number recitals as his or her reason for bringing up a safety concern. "You can't do it that way because OSHA 1910.134 says so" is one example. A worst case is the FDISO who tries to wash his hands of a safety infraction. Usually, this action is a result of an FDISO who is unable to get an incident commander or other officer to change an operation or task.

In all cases, the FDISO who uses CYA tactics is destined to fail. Firefighting crews will see right through any interventions and simply dismiss the FDISO recommendations or orders as pure self-preservation. To avoid this, the FDISO must display a genuine concern for everyone's safety. He or she must take personal responsibility for every firefighter's safety. Although codes and regulations may tell us safe ways of doing things, the reason we do things according to the document is not because of the document, but rather because the FDISO believes in doing things the safe way to reduce the threat of injury. This approach is best labeled "good intent" and "personal concern."

The Worker

The worker trap is one where the FDISO pitches in and helps crews accomplish their task (Figure 5-6). To be effective, the FDISO must stay mobile and not allow him- or herself to get trapped into assisting with a fixed assignment. Too often, this is easier said than done. In many cases, the FDISO may find it easy to pitch in and help move hoselines or throw ladders because there simply are not enough people present to do all the assigned tasks. If this is the case, the incident commander should consider taking back the safety officer responsibility and allowing the former FDISO to lead an assigned crew. Or, and perhaps more effectively, the tasks and risks should be reevaluated and prioritized to match the number of persons available. If there are too few people to do assigned tasks, then the role of the FDISO becomes even more critical and the need to see the big picture and not get caught up in activities should be further resolved. Before his retirement, Division Chief Gene Chantler of the Poudre (Colorado) Fire Authority said it best when he said, "If an officer assigned as SO (safety officer) becomes directly

Figure 5-6 *The FDISO must be careful not to get sucked into performing tasks. (Photo courtesy of Richard W. Davis.)*

involved with suppression, there are two people who made a mistake. The first is the individual assigned to the position; it is clear they are unsuited to fill that role. The second person to make a serious mistake is the IC for appointing someone to the position who doesn't clearly understand what the job entails."[2]

These three traps can, and have, wrecked many incident safety officer programs. The FDISO must look past these traps and be constantly vigilant to avoid them.

INCIDENT COMMANDER RELATIONS

■ Note

Making a difference as an incident safety officer is codependent on an FDISO who supports the incident commander and an IC who believes in his or her FDISO.

Making a difference as an incident safety officer is codependent on an FDISO who supports the Incident Commander and an IC who believes in his or her FDISO. Together, injury and death potential will be reduced. In order to achieve this harmony, the FDISO must strive to present tangible, well-articulated hazard observations. Further, the FDISO should steer clear of any statement, feedback, or action that subverts the IC. In order to avoid trouble with the IC and work toward a safe incident scene, the FDISO should embrace the following key points.

Authority

In virtually all command systems, the Incident Commander is in charge and the Safety Officer is a direct report to the IC, which means that the Incident

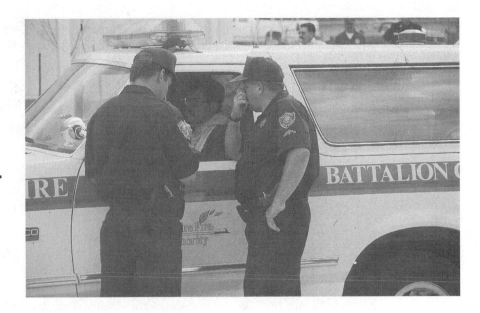

Figure 5-7 *The ultimate authority for firefighter safety rests with the incident commander. The FDISO must respect this.*

Commander is ultimately responsible for the safety of every person operating on the incident scene (Figure 5-7). Viewed simply, the FDISO should yield this authority and not pursue any argumentive approach to correcting tactics or strategies that the IC disagrees.

If the FDISO feels strongly about asking the Incident Commander to change a situation, the FDISO has the responsibility to speak clearly, professionally, and rationally to the IC. Once the IC has shared his decision, the FDISO must accept the decision and move forward. To belabor the point can be damaging to relationships and does not help remove the concern. Perhaps the FDISO can return to information gathering or offer peripheral solutions to address the hazard in question. Just as the fire or incident progression is dynamic, so should be the IC–FDISO relationship. Perhaps the situation will change and the FDISO suggestion becomes valid to the IC. Remember, the vision or outlook of the IC and the FDISO is similar, yet different. Perhaps the best way to say this is to quote a passage in the U.S. Fire Administration's Risk Management Practices manual:

> Incident commander's view:
> "Get the job done and operate safely."
>
> Incident safety officer's view:
> "Operate safely and still get the job done."[3]

Communication

Most multi-firefighter fatalities are blamed on communications (or lack thereof). Using all communications tools available, the FDISO should maintain contact

Figure 5-8 *The FDISO can be part of the solution path, as opposed to just bringing problems to the table.*

with the IC. Although radio communication is often essential, the face-to-face method is most effective in communicating with the IC. The face-to-face method allows for dialogue and feedback and mostly eliminates the barriers found with radio transmissions. Further, face-to-face allows both parties the opportunity to see if the message is understood. As a rule, communication should take place at least every 15 minutes on routine incidents and sooner if conditions or factors change. Likewise, if the FDISO has intervened with any assignment of any work crew, the IC must be immediately informed. Items that should be communicated often include fire progress, structural type and integrity, known hazards, potential hazards, and task effectiveness.

Solution Driven

Problem solving is not only required, but expected of an incident commander. Too often, subordinate officers bring problems to the IC to be solved. While the authority and responsibility to solve these problems rests with the IC, it seems reasonable to bring the problem AND a solution to the IC for consideration (Figure 5-8). This accomplishes two things: First, the IC gets a head start on problem solving, that is, one solution is already drafted and second, the IC can troubleshoot the solution and offer guidance to a future IC.

The FDISO can embrace this process both as one who brings problems to the IC and as one who receives safety concerns from working crews. In doing so, the Incident Commander looks at the FDISO as a partner and the FDISO extends credibility and safety awareness behaviors to the crews.

Figure 5-9 *The FDISO should strive to be a true "consultant" for the incident commander.*

Consultant

The most effective incident safety officers are those who operate as a consultant (Figure 5-9) to the incident commander. The FDISO operates like a technical consultant who asks questions, defines problems and strengths, drafts solutions, and offers recommendations to those who make decisions. Specifically, the FDISO establishes a relationship with the incident commander by asking what the action plan is, followed by a summary of the current situation status and resource status. With this information, the FDISO can collect more information in the form of a reconnaissance or 360° size-up of the incident. With this additional information, the FDISO can report concerns and possible solutions to the IC. The IC may direct further focus or evaluation. Along the way, the true consultant offers small suggestions to the working elements that can help them do their job smarter or safer.

As you can see, the FDISO who works as a true consultant is apt to make a difference with working crews and with smooth relationships with the Incident Commander.

Summary

The technically competent Incident Safety Officer is only as effective as his or her ability to trigger safe behavior. Further, the FDISO should strive to avoid common traps that can render the FDISO ineffective. Methods to create a safe incident range from the subtle to the firm. In most cases, a simple, nonthreatening reminder or awareness statement is all that is needed to correct an unsafe situation. In life-threatening situations, the FDISO should move toward firm intervention. It is important to avoid common traps that keep the FDISO from seeing the bigger picture of incident safety. Focus on PPE is best handled at the company level as opposed to the FDISO. Also, the FDISO who approaches his or her tasks with a self-liability protection view is likely to be ineffective. The FDISO's relationship with incident commanders should be one of a positive, solution-driven consultant that is neither argumentative or abusive of authority.

Review Questions

1. List four common triggers that lead to improved incident safety.

2. Give three examples of FDISO traps that can hurt FDISO effectiveness.

3. List and explain four components that will improve incident commander relations.

Notes

1. A. Schweitzer, *Great Quotes from Great Leaders* (Lombard, IL: Great Quotations, 1990), p. 31.

2. Excerpt from "The Safety Officer: A Roundtable," *Fire Chief Magazine* 3, no. 2, (February, 1993), p. 34. Communication Channels, Inc., Atlanta, GA.

3. *Risk Management Practices for the Fire Service,* United States Fire Administration, FA-166, Emmitsburg, MD: (1996), p. 78.

2

ON-SCENE EFFECTIVENESS

So far this text has looked at building a solid foundation for preparing the Incident Safety Officer. Further, we have looked at ways to develop a system that ensures that an incident safety officer is available for assignment. We also identified some triggers to spark safety consciousness on scene as well as some common traps that thwart good safety intentions.

Now it is time to put ourselves on scene and use a mental processing model—the Incident Safety Officer Action Model—that helps the FDISO juggle the many responsibilities, duties, observations, evaluations, and judgments that an Incident Safety Officer must perform in order to be effective. Further, this section discusses the postincident responsibilities of the FDISO and presents a simple, effective method to start an investigation following a firefighter injury.

Chapter

6

The Incident Safety Officer Action Model

Learning Objectives

Upon completion of this chapter, you should be able to:

- Describe typical duty frustrations experienced by the FDISO.
- List the four components of the FDISO Action Model.
- Explain the importance of cyclic thinking in the FDISO role.

The sudden vibration of your digital pager causes you to jump—you're still not used to wearing the new device, just as you feel a bit nervous and excited about being the newest member of your fire department's duty safety officer program. You look at the alpha readout and it says:

STRUCTURE FIRE 900 S WILSON. FIRST-DUE REPORTS HEAVY FIRE SHOWING, SECOND ALARM REQUESTED. CONFIRM THIS PAGE. –DISPATCH

You try to calm yourself knowing that you've got 12 years of experience and have seen many fires. Yet being "Safety" is different. The questions flood through your mind. Will I be able to get a good building read? Will I remember everything I need to look for? Were did I put my checklist? What's our department's risk plan again? Will I be able to prioritize multiple hazards? Can I protect our people?

DEVELOPMENT OF THE ACTION MODEL

So far, this book has focused on preparing to operate as a fire department incident safety officer (FDISO). The focus now shifts toward fulfilling the incident safety officer assignment. Without a systematic approach to accomplish all the mental and physical tasks that the FDISO is responsible for, the FDISO is destined to mediocrity. One particularly effective systematic approach to performing as an incident safety officer is to use the Incident Safety Officer Action Model, which was developed to address a central frustration expressed by fire officers wishing to better perform as a FDISO. Let us look at the root of this frustration.

■ **Note**
Performing as an FDISO can be viewed as the quintessential reactionary as well as the idealistic proactionary.

Performing as an FDISO can be viewed as the quintessential reactionary as well as the idealistic proactionary. On one hand, the incident safety officer must look at what has already happened and offer solutions to correct unacceptable risk situations. On the other hand, the FDISO has the opportunity to predict future events and make suggestions to minimize the effect of those events on the firefighter. Which takes priority? Obvious arguments can be made either way.

Another source of frustration to being an effective FDISO has to do with the prioritization of physical and mental tasks that the FDISO must address. Often, an FDISO arrives at the incident scene after initial fire attack or setup. After checking in with the incident commander, the FDISO hopes to get a quick briefing on what has happened, what is planned, and what is needed by the IC. What is next? Often a 360° scene survey is requested by the IC, other times the IC may have a specific question or concern he or she wants addressed (like, "When is this building going to collapse?"). Other times, the IC leaves the options up to the FDISO by saying, "I don't have the full picture here. Find out what the heck is going on!" Worse yet, the IC may just assume that you, the FDISO, know what needs to be done and give no indication of an assignment. It is not at all uncommon for an IC to have the FDISO draft a quick action plan based on the current

> **The Incident Safety Officer:**
>
> Quintessential   Idealistic
> Reactionary   or   Proactionary
>   ?   ?

and predicted situation status and resource status. As you can see, it is hard to develop a starting place for the many items that need to be addressed by an Incident Safety Officer, leading to frustration.

In the early 1990s, approximately 500 feedback and review sheets from the Fire Department Safety Officer's Association course, Preparing the Fireground Safety Officer, were reviewed. When asked, "What can we add to the class to better address your needs?" respondents listed numerous frustrations and asked for tools to address these. A sampling of these frustrations include:

- There are no clear starting places that can be assumed for FDISO duties.
- Existing Incident Safety Officer checklists are too short and/or too general to provide what is needed to be effective.
- A checklist that includes everything that an FDISO should address would be very long and unrealistic to apply to every incident.
- The FDISO must be reactive to the needs of the Incident Commander as well as proactive in the prevention of injuries to firefighters. These create a priority clash at times.
- The FDISO must stay flexible and not be sidetracked with details that can obscure the big picture. How do we do this?
- We need a strict and reliable method to take immediate action whenever the situation warrants.
- An expectation exists that the FDISO needs to see all and know all in order to be effective. We need a tool to help us manage this.
- Typical checklists imply that once an item is checked off, it no longer needs attention. This could be a dangerous assumption for an FDISO.

With these issues in mind, this author and Battalion Chief Terry Vavra of the Lisle-Woodridge Fire District met at the Fire Department Safety Officers Association office in Ashland, Massachusetts, to plot out a systematic method to teach, track, and apply incident safety officer duties. In that meeting, each issue and concern was evaluated with an eye toward developing a model to mentally process the various facets of an incident. Experiences from around the country, course critique suggestions, base information from the original Preparing the Fireground

Incident Safety Officer course, and professional visitations to large and small fire departments became the data from which a needs list was developed. It was believed that a simple, easy-to-apply model could be created that was adaptable to most incidents, overcoming previous frustrations. Interjected here was the need to create a mental process that we call "cyclic."

Using the mental process of cyclic thinking, the needs list, and our own experience as incident safety and training officers, we created the Incident Safety Officer Action Model.

ACTION MODEL COMPONENTS

FDISO Action Model
a four-area thought-processing model that assists the incident safety officer in remembering items that must be evaluated during emergency incidents

The **Incident Safety Officer Action Model** is a cyclic, four-area model that allows the incident safety officer to mentally process the surveying and monitoring of typical incident activities and concerns. As shown in Figure 6-1, the model has four general areas that need to be addressed by the FDISO. No starting place or direction of flow for the model should be implied. Upon check-in with the incident commander, a starting place may be assigned. If so, the FDISO simply jumps onto the cycle as directed. If the IC does not direct the FDISO to a starting place, the FDISO can start where he or she feels attention is warranted. Once on the cycle, the FDISO should conduct an initial survey of each area, then monitor the applicable concerns within each. Recognizing that the FDISO must stay flexible to the incident and the incident commander's needs, no direction is established. In essence, the FDISO performs a mental evaluation of the conditions, activities, operations, or probabilities within each area. A general overview of each model area is presented here. Subsequent chapters explore each area with specific applications, concerns, warnings, examples, and explanations.

Of particular importance is the cyclic design of the Incident Safety Officer Action Model (Figure 6-1). All of us use a linear thinking process to handle incidents, that is, having a defined starting point with a desired end-point. During the thinking process, inputs are made along the linear path. A point may be arrived at where the person making decisions has reached maximum input, causing stress. Combine this point of maximum input with the focus to reach an ending point and it is easy to see that some hazards may be overlooked. This point was highlighted following the tragic Storm King Mountain fatalities during the South Canyon Fire in Colorado.

Fourteen firefighters died in a sudden blowup of the South Canyon Fire in 1994. Following the incident and subsequent investigations, human error and communication breakdown were cited as contributors to the incident. Ted Putnam, Ph.D., specialist for the Forest Service's Missoula Technology and Development Center wrote an article entitled "Collapse of Decision Making" which appears in the publication, *Findings from the Wildland Firefighters Workshop*[1] that discusses the psychological elements that lead to failed leadership during wildland incidents. In the article, Dr. Putnam makes a case for the traps of linear

THE FIRE DEPARTMENT INCIDENT SAFETY OFFICER ACTION MODEL

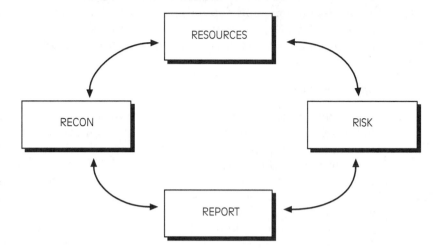

Figure 6-1 *The four Rs of the Incident Safety Officer Action Model: Reconnaissance, Resources, Risk, and Report.*

thinking by discussing decision-making models and conclusions based on numerous studies. Dr. Putnam concludes that increasing incident stress actually leads the decision maker to minimize the number of inputs being considered and the person regresses toward a more habituated behavior.

It is imperative that a decision maker creates an environment where he or she can stay open to multiple inputs and maintain a high degree of situation scanning and awareness. To help accomplish this, the IC can increase the number of people working closely with him or her—thereby increasing the number of inputs that can be processed. The FDISO can help the IC with situational scanning and help increase the number of inputs that the IC can receive—namely, hazard recognition and safety concern inputs. Further, the FDISO can heighten his or her own environmental scanning ability by using a mental image that is circular—a reminder to stay open to the hazards that may eventually cause injury to firefighters. A system of cyclic, or recurring evaluation by the FDISO can help eliminate the trap of underestimating hazards.

The Incident Safety Officer Action Model uses a cyclic, visual pattern to remind the FDISO of the need to constantly reevaluate each component of the model. Briefly, the components requiring evaluation and attention are mentioned here and in subsequent chapters. To make each area easy to remember, you can think of the Action Model components as the *four Rs*.

■ Note

A system of cyclic or recurring evaluation by the FDISO can help eliminate the trap of underestimating hazards.

Resources

It is easy to say that there are *never* enough people or equipment to handle a significant incident. Incident commanders earn their respect by their ability to manage the resources that are available. Given this, you, the FDISO, must evaluate the resources on hand and determine if those resources match or support the action plan. Specifically, the FDISO needs to evaluate the resources of *time, personnel,* and *equipment.* These are discussed in depth in Chapter 7.

Reconnaissance

Most incident commanders agree that having an Incident Safety Officer available to help do a 360° scene survey is essential to improving scene safety. This survey is often called *reconnaissance* (or **recon**), an exploratory examination of the scene environment and operations. The effective FDISO uses this recon trip to evaluate numerous facets of the situation from building characteristics, to fire conditions, to crew effectiveness, to fall hazards. For the sake of this book, reconnaissance has been divided into two categories, environmental survey and operational survey. Chapter 8 deals with the environmental issues (building integrity, access, weather, etc.) that need to be evaluated and reported to the IC. Chapter 9 looks in depth at the evaluation of the operational issues (tools, tasks, teams, etc.) that may produce injuries for fire fighting crews.

recon
shortening of the word reconnaissance; seeking out of critical information

Risk

Chapter 2 addressed the concepts of risk management. On scene, these concepts must be put into practice by the Incident Commander through the observations and efforts of the Incident Safety Officer. Chapter 10 outlines a process for determining, on scene, acceptable and unacceptable risk.

Report

What seems like an obvious responsibility of the Incident Safety Officer is actually one of the most forgotten. Many of the conflicts that exist between incident commanders and incident safety officers can be resolved through timely, appropriate updates. Additionally, the report phase of the action model reminds the FDISO that written reports, safety briefings, and review of any action plans are necessary. Chapter 11 presents an easy-to-follow process for reporting conditions, reviewing action plans, and writing safety briefings.

By addressing each of the four *Rs* in a continuous, cyclic manner, the FDISO can improve the safety of a given incident. Further, the model can help reduce FDISO frustration and improve communications with incident commanders. The remainder of this book, explains each component and notes specific items to look for and act on to improve firefighter safety thereby *making a difference* as an Incident Safety Officer.

Summary

The incident safety officer can be filled with frustration due to the expansive and dynamic nature of the job. Further, the FDISO's focus to catch unsafe conditions before injury occurs may well be distracted as the incident grows in complexity. The use of the Incident Safety Officer Action Model will help the FDISO juggle the dynamics of a typical incident scene and provide a cyclic-thinking model to keep the FDISO focused on hazard recognition and abatement. The cyclic-thinking feature is instrumental in that it keeps the FDISO from a linear or checklist approach that may have the effect of minimizing hazards or not returning to issues to reevaluate.

Review Questions

1. Briefly explain common frustrations that can distract the FDISO.
2. List the four components of the ISO Action Model.
3. Explain the concept of cyclic thinking. Explain the importance of cyclic thinking for the FDISO.

Note

1. T. Putnam, "Findings from the Wildland Firefighters Workshop," USDA Forest Service Publication, #9551-2855 MTDC, (Missoula, MT, July 1996).

Chapter

7

Incident Resource Evaluation

Learning Objectives

Upon completion of this chapter, you should be able to:

- List the three resources that need to be evaluated by the FDISO.
- Describe the relevance of time on an incident and list methods to achieve more accurate timekeeping.
- Apply a sense of appropriateness of the number of people an incident requires.
- Define *risk-exposed personnel*.
- List various accountability systems as well as advantages and disadvantages of each.
- Discuss equipment as a resource.
- List common incident hazards and define a method of protection from each.

The victim has been trapped for over an hour now. Quick rescue techniques have not been successful in getting the alive victim out of the 20-foot deep shaft. The Incident Commander, you the Safety Officer, and the extrication team leader have just agreed to an action plan that involves the digging of a parallel shaft and cross tunnel to reach the victim. The Logistics Section Chief is busy ordering equipment and the extrication team leader is briefing his crews. The IC steps over to you and asks in his typical light style, "OK Mr. Safety-kinda' guy, what are we missing? What do we still need to address?" Instantly, you think of the hazards created by just the sheer resources of people, equipment, and extended incident duration. You reply with a smile. "OK Mr. Incident Commander-kinda' guy, let me outline just a couple of hazards we need to be watching for."

The Incident Safety Officer Action Model can help the fire department incident safety officer (FDISO) mentally process the many factors that determine scene safety. This chapter covers one component of the action model—resources (Figure 7-1).

The ability to evaluate, manage, and assign resources can become the hallmark of an exceptional Incident Commander. Budgetary concerns, response times, politics, and compatibility issues combine to limit the availability of resources to overcome a given incident. Although it would be nice to have unlimited resources to handle a serious incident, it is unrealistic to expect. The FDISO must look at what *is* available and determine if adequacy exists. Many types of resources need to be evaluated but generally, time, personnel, and equipment are

THE FIRE DEPARTMENT INCIDENT SAFETY OFFICER ACTION MODEL

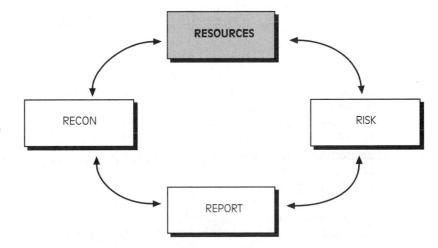

Figure 7-1 *The "Resource" box of the Action Model reminds the FDISO to evaluate time, personnel, and equipment.*

key considerations. The basic question the FDISO must ask with each of these is, "Do we have enough to do what we're trying to do?" Here, we review each area and look at specifics that help answer this question.

TIME AS A RESOURCE

Time is often forgotten as a resource. Often, time is considered a size-up factor. When looked at from the FDISO perspective, time is an important resource that dictates hazard priorities, reconnaissance efforts, or makes the difference between imminent threat to life or minor operational concern. The FDISO must have a keen sense of time. This may sound easy because, after all, time is constant. Time cannot speed up, it cannot slow down. You cannot trade it or buy it. You cannot keep it or give it away. Time does, however, tend to "slip away" from time to time (pun intended!). Simply said, our perception of time changes.

Most experienced fire officers can recall a situation where a fast, aggressive fire attack seemed like a 10-minute accomplishment when, in reality, it took 30 minutes. Conversely, those same officers have experienced the life-saving effort that seemed like it took 30 minutes, only to find out that 10 minutes elapsed. The FDISO CANNOT allow this "warp" of time. Why? Frankly, lives depend on it. It is not hard to imagine an FDISO, concerned about a collapse potential, getting wrapped up in a building preplan book trying to determine the building's exact roof structure, only to miss the rapid extension of the fire into the rest of the structure and the signs of sure collapse.

How can the FDISO evaluate and use time effectively? One suggestion is to preplan the use of a dispatch or communication center to assist (Figure 7-2). The assistance comes in the form of a general broadcast of transpired incident duration over the assigned incident radio frequency. It would be great for the FDISO if a communications center announced time intervals:

"Fourth Street Command, Dispatch. You are 15 minutes into your incident."

"Fourth Street Command, Dispatch. You are 30 minutes into your incident."

Computer-aided dispatch centers can usually accommodate this request with a built-in timer that beeps or signals the communications technician that an update is needed.

More simply, a wind-up egg timer can be used by the dispatcher; a ding means it is time to announce the incident duration. Fifteen minutes seems to be a reasonable time duration reminder although some departments have chosen 30. Many departments have chosen 15-minute announcements as it easily correlates to self-contained breathing apparatus (SCBA) bottle use, additional resource response times, rehab needs, fire propagation, and provides a good pace to make reports to the IC. Palm Beach County Fire and Rescue (Florida) created a two-tier timekeeping sequence as a postincident analysis recommendation following their well-documented Builder's Square Fire in 1993.[1] In a two-tier system, the dispatch center notifies the Incident Commander every 5 minutes for the first 20 minutes during working incidents (incident plus 5, incident plus 10, etc.). After

■ **Note**

Time cannot speed up, it cannot slow down. You cannot trade it or buy it. You cannot keep it or give it away.

Figure 7-2 *Dispatch centers can help time management by announcing time in 15–30 minute intervals.*

20 minutes, the timekeeping moves to a 15-minute interval. Hopefully, the FDISO can use a duration reminder to help pace the incident. The Pittsburgh, Pennsylvania, Fire Department initiated this type of timekeeping following a triple-firefighter-fatality fire in a single-family dwelling.[2]

If a timekeeping service is not available through a dispatch center, the FDISO should use a stopwatch, egg timer, or other device to help monitor the passage of time.

Once the FDISO has command of time as a resource, better judgments can be made regarding prioritizing duties, technical considerations, and overall scene risk. The following are examples of situations where time becomes a resource for the FDISO:

Response Time

The FDISO may find a hazard that indicates the need for more engine or truck companies if the action plan is to work. Knowledge of the response times of these resources becomes essential to determine if the response will make a difference. Many times, the time it takes to get additional apparatus and crews in place exceeds the window of opportunity—the fire will simply get too big, the victim may die, or the building will collapse. This is especially true for rural areas where a dependence on mutual aid companies exists.

Response time also applies to other resources requiring the attention of the FDISO. A perfect example of this is the request and response of medical evacuation helicopters. Many departments have embraced the helicopter as a life-saving answer to long transport times. Once a medical evacuation helicopter is ordered, the FDISO needs to monitor time and make sure that he or she has evaluated the designated landing zone (LZ) prior to the time of helicopter arrival (Figure 7-3).

One note of clarification here: It is *not* suggested that the FDISO be assigned the task of establishing an LZ or ground contact for incoming helicopters. It is suggested, however, that the FDISO evaluate the LZ and verify that ground personnel have suitable cover in the event of a mishap with the helicopter. Dick Silva, an experienced and celebrated television network helicopter pilot, once told a group of firefighters, "Tell me where you want me to land, then get the hell outta there. Have you ever seen a helicopter rollover? You don't want to be anywhere near the thing!"

Projected On-Scene Time

When looking at the big picture, the FDISO needs to project the on-scene time. Obviously, a 2-hour firefighting effort requires a different crew pace than a 2-day collapse rescue. The FDISO can help set up crew pacing given this projection. Once an on-scene time is projected, the FDISO can better offer judgment and recommendations on the need for, and scope of, a crew rehabilitation program. Short duration incidents (less than 2 hours) may require just rest, health monitoring, and hydration. Longer incidents may indicate the need for nourishment, extended rest, or fresh crews. Rehabilitation is covered more specifically in Chapter 9.

Figure 7-3 *The FDISO should evaluate the helicopter LZ to ensure that personnel will not be endangered if a mishap were to occur.*

Action Time

For any tactical assignment, a given amount of time is required to accept, initiate, and accomplish the task. We call this *action time*. The incident commander orders tasks done because a window of opportunity exists where completion of the task will benefit the entire operation. The FDISO must have a strong sense of action time for many scenarios. As you can imagine, many variables affect action time. These might include:

- Number of firefighters on the crew
- Training level of crew members
- Availability of equipment to perform task
- Access to area where tasks are to be performed
- Restrictive factors that impede task completion (smoke, weather)

The FDISO should evaluate the progress of the task as it relates to time and be prepared to offer judgment on whether the task can be completed in the opportunity window and before another event creates conflict. One important example is the timing of a roof ventilation assignment and potential collapse. The FDISO must be aware of the time it is taking a crew to open up the roof and compare this to the type of roof construction and location of the fire. The FDISO may know that the crew assigned to the roof is using a chain saw and that, based on previous fires and training evolutions, it should take them 8 minutes to have the roof open. If 12 minutes have passed and still no signs of ventilation have appeared, the FDISO is prudent to check with the crew and the IC and to relay concern about fire spread, structural integrity, and so forth. In cases like this, the window of ventilation effectiveness by a firefighting crew may have passed, especially in this era of lightweight wood truss, meaning self-vent from collapse or burn-through. In cases like this, an eminent life-threat may exist—the FDISO must act to evacuate crews prior to collapse!

PERSONNEL AS A RESOURCE

Without a doubt, personnel are the most important resource to consider when evaluating scene safety. For the FDISO, the evaluation of personnel safety takes on many forms—this entire book is dedicated to that premise. For discussion here, however, the FDISO needs to focus on evaluating three key components relative to personnel: total scene attendance, high-risk exposed personnel, and effectiveness of an accountability system.

Total Scene Attendance

The basic question here involves a judgment on how many people are necessary to handle this incident. Are there enough? Each department struggles with the basic issue of getting enough people to the scene to handle the problem. Some departments, although rare, have to deal with too many people at an incident.

Figure 7-4 *The basic question should be asked: "Are there enough people on scene to do what is required?"*

Many studies have examined the typical fire scenario and offered suggestions on the number of people required to effectively handle the incident. Most of these studies have focused on the initial response in order to help establish minimum staffing for initiating firefighting. For the FDISO, the number of people needed to handle a given incident must be based on understanding the action plan set in motion by the incident commander and the potential duration of the incident (Figure 7-4).

The following scenarios suggest total scene attendance required from the FDISO perspective. These numbers are meant to show one perspective; they are merely a starting point for planning and debate.

Situation 1

Incident: Single-family, single-story dwelling, two-rooms and contents burning.

Action Plan: Simultaneous fire attack with one crew, search with another, roof ventilation with another.

Projected Incident Duration: 1.25 hours

Total on Scene Required:

3	Fire Attack
2	Search
3	Roof Ventilation (1 on the ground)
3	Rapid Intervention Team
4	Staged (backup/rehab/SCBA change)
2	Pump Operators
1	Incident Commander
1	Scene Safety Officer
19	Total

FDISO Perspective: The rapid intervention team is the immediate coverage for the two inside crews. Two additional crews can be rotated into the operation from staging. This could help keep crews on a two-bottle rotation pace.

Situation 2

Incident: Child trapped in 20-foot deep shaft

Action Plan: Maintain visual contact and air supply through cameras and pumped oxygen. Dig parallel shaft through mineral earth, encase/shore as necessary, dig connecting shaft. Provide for rapid medical needs and crew rehab.

Projected Incident Duration: 20 hours

Total on Scene Required:

2*	Victim monitoring
1*	Lead digger
1*	Lead digger air/rope monitor
4*	Dirt removal/shuttle crew
4*	Encasement/shore building crew
12*	Staged (for crew rotation)
2	Rapid Intervention Team
2	Medical standby and rehab
1	Equipment Operator (air)
1	Equipment Operator (lighting)
1	Logistics Section Chief
1	Incident Commander

1	Operations Section Chief
1	Scene Safety Officer
22	Total

* NOTE: These persons will set up a timed crew rotation. Staged persons are a one-for-one replacement for these crews. This staffing is for any one time period—a 20-hour incident may require the rotation of more than fifty people into the incident.

FDISO Perspective: Enough personnel must be available to set up a rotation system that allows long-duration operations. A 15-minute rotation may work to help prevent tunnel vision, emotional attachment, or fatigue. Medical standby and rehab are key to observing crews for the aforementioned conditions and syndromes.

One last note when considering total scene attendance. The FDISO is typically on scene for significant emergencies where many crews may rotate in and out of the incident. The FDISO should question if adequate facilities are available for human needs. Early suggestions for restrooms, sheltering, water, and so forth may need to be made.

High-Risk Exposed Personnel

Firefighters who have entered a structure for an aggressive interior firefight deserve extra attention by the FDISO. Typically, the FDISO cannot see exactly where interior committed firefighters are. Further, the FDISO may not know how many persons are committed. If this is the case, the FDISO needs to check briefly with the accountability system monitor or with the Incident Commander to ascertain this information. The FDISO should ask, "Who is at greatest risk here?" or "Given these conditions, who is most likely to be injured?" (Figure 7-5).

The basic idea here is to give the FDISO a starting place for evaluating risk and benefit. Solid risk management is actually a five-step process that includes hazard evaluation and prioritization (Chapter 2). In these steps, a determination is made as to frequency potential and severity of the consequences. Persons engaged in aggressive interior firefighting are in an environment immediately dangerous to life and health **(IDLH)** which can be classified as "high-frequency, high-severity" in most cases. In other words, those making an aggressive interior attack on a hostile interior fire are not only likely to be injured, but the injury will be severe. Given that, these firefighters deserve a high priority in addressing their needs or minimizing hazards. The interior-committed crew may be an obvious example of the risk exposed. The following are other examples of **hot zone** committed crews that require extra attention by the FDISO.

IDLH
immediately dangerous to life and health

hot zone
the area immediately surrounding and including an environment that is immediately dangerous to life and health

Incident Type	High-Risk Exposed
Motor vehicle accident	Traffic controller, those "between" vehicles
Mass casualty	Triage team

Figure 7-5 *The FDISO should pay particular attention to crews assigned to IDLH atmospheres.*

HazMat	Team in hot zone, decontamination support crew
Building collapse	Rescue/search teams
Wildland fire	Crews operating at the head of the fire or those working in the path of the fire

accountability system
a system that readily identifies the location and function/ assignment of all personnel operating at an incident

Accountability Systems

NFPA 1500, Standard on Fire Department Occupational Safety and Health mandates the use of an **accountability system** on all emergency scenes.[3] The same standard gives the local jurisdiction the ability to design this system to meet the needs of the locale (Figure 7-6), so many different forms of accountability have been developed around the country. Some of the more common systems are:

Figure 7-6 *The FDISO should check to see that the accountability system is working effectively. (Photo courtesy of Richard W. Davis.)*

Passport This name is given to a crew-card system that was primarily designed in Seattle and Phoenix. In essence, a company officer carries a card and each member assigned to that crew puts his or her Passport on the card. The officer is responsible for checking into the incident and the crew's card is tracked on a status board. This system has reportedly been successful with many career departments.

Cow Tag This type of system includes any system where individual members have a tag with their name, equipment number, or other identification assigned to them. The tags can be metal, cloth with Velcro, or laminated paper on a ring. Upon arrival on scene, the member places his or her tag on a ring on an apparatus or hands the tag to a staging manager. Someone must collect the tags or arrange the tags into crew assignments to track members. This system is popular in volunteer, combination, and paid-on-call departments.

Company Officer This system relies on a company officer to track his or her assigned members. The Company Officer is responsible to inform the Incident Commander of assignment status. To be successful, this system requires strict discipline of crew members and the company officer as well as a well-tracked status board. This system is prevalent in career departments that have chosen not to utilize devices for accountability.

Bar Code This high-tech method can process either a tag system or a passport system. A bar code scanner is used to track units on an electronic or computerized status board. This system is currently used by the Fairfax County (Virginia) Fire and Rescue Department.

■ Note
It is not suggested that the FDISO manage the accountability system: The system should be automatically used on every scene.

Each of these systems has advantages and disadvantages that the FDISO needs to understand so that problems with its use on scene can be overcome. Figure 7-7 lists some desirable traits of an accountability system.

Many departments have adjusted their accountability system so that a person who has not checked-in will be obvious. One Pennsylvania fire chief was frustrated that his firefighters were not turning in their tags so he had hangers placed on everyone's helmet so that the tag would hang in front of their face!

For an accountability system to be effective, the incident commander or FDISO should be able to tell how many crews are assigned tasks, what those tasks are, and the crew's probable location. Also, the system should account for crews available in staging and those assigned to rehab. If this does not exist with the system the FDISO has to work with, then the job of tracking firefighters must be assigned and these points covered immediately. This action can be followed up by a strong FDISO recommendation that the department change its program to incorporate these points. Some departments may view this change as unnecessary or too complex; those departments must ask themselves, "How fast do we know that everyone is out when we order an emergency evacuation of the building? Who is missing if the numbers do not add up? Where should I start my rescue effort?"

With all accountability systems, the FDISO needs to ensure and enforce that the system is being used as intended. One progressive North Carolina fire chief proposes discipline for members who do not participate in the on-scene accountability system—up to and including time off.

It is not suggested that the FDISO manage the accountability system: The system should be automatically used on every scene. If the FDISO is tied down to initiating and updating the accountability system, then other essential FDISO duties cannot be accomplished.

Accountability systems are not a replacement for self-disciplined crews or wise crew supervision. The hallmark of accountability is the ability of all

- Utilized at *all* incidents
- Simple to initiate and maintain
- Can instantly provide information as to crew make-up, assignment, and location
- Accounts for persons in staging and rehab
- Tracks assignment changes in real-time

Figure 7-7 *The elements of an effective accountability system.*

!Safety
● Crew officers MUST be able to account for the location and actions of their members. The FDISO needs to support this concept when evaluating the accountability of crews.

■ **Note**
As with the incident commander, the FDISO should be tracked by the accountability system.

firefighters to stick with their assigned crew and to follow the instructions of the crew officer. Crew officers MUST be in a position to account for the location and actions of their members. The FDISO needs to support this concept when evaluating the accountability of crews. If a person is found to be freelancing, the individual should be approached and directed to a crew. In addition, a crew officer needs to be assigned the responsibility to keep this from happening again.

Tracking the FDISO   More than once, an FDISO has been criticized for not having a partner or not being tracked by the accountability system. As with the incident commander, the FDISO should be tracked by the accountability system. The FDISO should act as an example, if for no other reason. When performing a roving assignment or entering a collapse or hot zone, the FDISO should have an assigned partner, just as we expect firefighters to work in teams or pairs. This is especially important when the IC wishes the FDISO to get a close-up look at a fire floor, collapse zone, or other assignment where a certain risk is taken to collect information. Failure to track the FDISO, or worse yet, failure of the FDISO to work as a team member in critical areas, will do grave damage to the safety efforts of the department, not to mention increase the chance of injury or death to the FDISO.

In one example of tracking the FDISO, Loveland Fire and Rescue responded to a nighttime incident where demolition workers started a small fire on the fifth floor of an old sugar-processing factory. Many of the ovens, towers, piping, stairs, and structural elements had already been removed, leaving huge voids and drop-offs. The incident commander took a nonaggressive approach and ordered all crews to work at half speed and only if they had plenty of light, even at the risk of the fire spreading. The assigned FDISO was instructed to go to the fifth floor and establish safe zones and to barricade hazards. Upon receiving the assignment, the FDISO asked the IC to assign another firefighter to the FDISO and to label the team "Safety Group." Additionally, an overall outside FDISO was requested because the original FDISO was actually assigned a task and a crew. At first the IC didn't understand the request, but after some thought, bought into the example this approach would create. This potentially deadly situation was handled without injury.

EQUIPMENT AS A RESOURCE

More often than not, the FDISO is concerned with not having enough people to safely mitigate the incident. Rarely is equipment a resource to evaluate. There are, however, equipment circumstances that need to be evaluated. For example, a standard assignment of two engines, a truck, and a rescue to a reported structure fire may seem appropriate, only to find out that the structure fire is a result of an aircraft crash that introduced 3,000 gallons of jet fuel into the situation! The FDISO can perform an equipment evaluation by categorizing equipment into two areas: apparatus and personal protective equipment (PPE).

Apparatus

What do the following have in common? Engine, Pumper, Tender, Tanker, Tank, Truck, Aerial, Tower, Squirt (Squrt), Quint, Quad, Squad, Rescue, Brush Patrol, Hammer, Hazmat, DECON, Dive, STAR, SWAT, Support, Salvage, Air, Air Tanker, AirLife, LifeStar, Life Guard, Guard, Flight, LifeFlight, Power, Deluge, C/O, Task Group, Strike Team, Rig, Red, CFR, Foam, ARFF, Paramedic, Ambulance, Car, Admin, Container, CV, SSV, Marine, Whale, Water, Wagon, and Unit." You're right: It's the vernacular emergency services uses to describe the many apparatus resources we employ to handle the 911 mission (Figure 7-8).

For the FDISO, particularly on mutual aid and expanded resource situations, the vernacular alone can cause significant confusion. Confusion may lead to an unknown hazard that needs to be uncovered. In many departments, the hazards and capabilities of Engine 1 are vastly different from the hazards and capabilities of Engine 3. Engine 1 may have a 2,000 gpm pump, 500 gallon tank, 65-foot telesqurt, and a 10Kw generator. Engine 3, on the other hand, may be only 1,250 gpm, hold 1,000 gallons of water, and is set up with a 1,000-foot progressive wildland hose pack.

For the FDISO, an evaluation of equipment needs and use is done for two reasons: First, to determine if enough of the right type apparatus is present, and second, if that apparatus is being utilized as designed or if another type of apparatus is more appropriate. In each case, a dialogue with the Incident Commander

Figure 7-8

Specialization in fire apparatus and equipment presents the question: "Do we have the right piece for the job?"

is essential to get the right apparatus to the right assignment. A word of caution however: This is an area where the FDISO may be viewed as overstepping his or her assignment, because it is the responsibility of the incident commander (or other delegated position) to evaluate the number and types of equipment needed to handle the situation. If the FDISO questions this responsibility, command breakdown may occur. The FDISO operates from a safety viewpoint, that is, with an eye toward eliminating hazards or injury potential based on the equipment assigned.

When evaluating the appropriateness of the number of apparatus, a few guidelines can help. First, and foremost, the FDISO needs to know the capabilities of the apparatus that the department utilizes or requests from outside agencies. Size, weight, capacity, length, and special features need to be available for quick reference. Usually, driver/operators and equipment engineers can get you this information. Some departments, because of the wide range of "engines" they use, have created a unit index for quick reference (Figure 7-9). This index is invaluable in determining if the apparatus assigned to a particular task is appropriate or creates added risk.

The FDISO is wise to have similar reference charts from neighboring agencies. As an example, one medical evacuation helicopter service uses a single-engine Bell JetRanger that requires a 60 foot × 60 foot landing zone (LZ). Another service uses a twin-engine Bell 222 that needs a 100 foot × 100 foot LZ. Likewise, "Anytown Engine 2" may be a 500 gallon, 1,000 gpm pumper and "Sometown Engine 2" may be a 500 gallon, 1,000 gpm pumper with a 50-foot ladder and nozzle boom.

Second, the FDISO needs to look at the current use of a given apparatus and make a brief determination if the use is appropriate. To help make this determination, consider the following:

Flow Needed The FDISO should have a solid figure for the fire flow needed for a given operation. Chapter 9 explores this subject a little further. The FDISO should then evaluate the pumping apparatus available, as well as the devices used to apply the water, and advise of any conflict. Once again it may appear that the FDISO is questioning the decisions made by the Incident Commander. The spirit in which the FDISO should approach this subject is grounded in the fact that firefighting personnel making an offensive fire attack without proper fire flow are in a position to be overrun by the fire. This *is* a RED FLAG that needs to be addressed by the FDISO.

Reach/Aerial Use Many types of elevated devices exist and each has its own set of operating parameters. The FDISO can evaluate these only if he or she has a quick reference of the capabilities of each. Also, the FDISO can relate some building construction knowledge here. An aerial device is really a cantilevered beam that distributes a load. Using this knowledge, the FDISO can offer judgment as to the effectiveness and safety of a device used for hoisting, rescue, bridging, or access.

Loveland Fire Department Vehicle Quick Reference Chart

* * * PUMPERS * * *

ID#	Year/Make/Model	Water	Pump	GPM	GVW	Eng/Trans	Features
Engine 2	'89 General/Spartan	1000	Waterous 2-Stage	1500	18.3 T	Diesel 6V/4 Auto	Top panel, 20 gal class A, stored class B
Engine 3	'96 Luverne/Freightliner	1000	Waterous 2-Stage	1500	20.5 T	Diesel 4S/5 Auto	PTO pump, 10k hyd-gen, 20 gal class A, progressive hose
Engine 22	'83 General	1000	Waterous Single	1500	18.2 T	Diesel 8V/4 Auto	PTO booster, top panel, 5.3K gen, stem light
Engine 33	'81 Sutphen	1000	Hale Single	1500	17.3 T	Diesel 8V/4 Auto	PTO booster, top panel, 5K gen, stem light
Engine 88	'92 General/International	1000	Waterous Three-stage	1000	17 T	Diesel 6V/4 Auto	4x4, pump & roll, 6K gen, 20 gal class A, progressive hose
Engine J	'98 General/Am. La France	750	Rosenbaur Single	1250	18T	Diesel 8V/4Auto	Rear mount pump, 10K gen, 20 gal class A, progressive hose

* * * SQUADS * * *

ID#	Year/Make/Model	Water	Pump	GPM	GVW	Eng/Trans	Features
Squad 1	'85 Spartan/Supervac	—	—	—	14 T	Diesel 6S/4 Auto	10-person cab, 12K gen, 2 light towers, hyd. PTO, cascade, heavy lift bags, heavy Halmatro set, shoring, confined space set
Squad 2	'84 Chev/Supervac	—	—	—	11 T	Diesel 8V/4 Auto	10K gen, 1 light tower, Hyd. PTO, cascade, full ground ladders
Squad 3	'80 Chev/Supervac	—	—	—	10.5 T	Diesel 8V/4 Auto	10K gen, 2 light towers, Hyd. PTO, cascade, rope rescue equip.

* * * AERIAL DEVICES * * *

ID#	Year/Make/Model	Water	Pump	GPM	GVW	Eng/Trans	Features
Truck 1	'75 Sutphen	300	Hale Single	1250	24.2 T	Diesel 8V/5 Man	85' Aerial w/Bucket & two 750 GPM nozzles, piped air to bucket, full ground ladders
Engine 1	'96 General/Spartan/Snorkel	750	Waterous 2-Stage	2000	25 T	Diesel 8V/4 Auto	65'Tele-Squrt, 1000 GPM auto-nozzle, 7.5K gen, 20 gal class A
Engine 4	'93 General/Spartan/Snorkel	750	Waterous 2-Stage	2000	25 T	Diesel 8V/4 Auto	65'Tele-Squrt, 1000 GPM auto-nozzle, 7.5K gen, 40 gal AFFF system

* * * SPECIALTY APPARATUS * * *

ID#	Year/Make/Model	Water	Pump	GPM	GVW	Eng/Trans	Features
Tender 1	'81 Supervac/Chev	1500	PTO Hale Single	350	14 T	Diesel 8V/5 Man	1500 gal porta-tank, manual turret, rear dump
Tender 3	'97 General/Freightliner	1800	PTO Waterous Single	500	18 T	Diesel 6S/4 Auto	1800 gal soft porta-tank, 20 gal class A, rear turret, 3-side auto dumps
Tender 9	'68 Kaiser	1000	Auxil Hale Single	300	15 T	Diesel 6S/5 Man	Forest Service Issue 6x6, 1000 gal porta tank
Rescue 4	'75 Walters	1500	Hale Single	800	16 T	Diesel 8V/4 Auto	ARFF Vehicle, 800/400 GPM roof turret, premix 3%, separate pump engine (diesel 6V), pump & roll
Rescue 44	'95 Walters/Ford	500	Darley Single	500	13 T	Diesel 8V/4 Auto	ARFF RIV, 500# dry chem, dual bumper turrets, 70 gal 3% AFFF, Halmatro, separate pump engine (diesel 4s); pump & roll, 7.5K gen w/own diesel engine
Dive Rescue	'75 Supervac/Chev	—	—	—	5 T	Gas 8V/3 Auto	All SCUBA gear, Zodiac Boat/outboard motor, Ice Suits, Dry Suits
HazMat	'87 GMC	—	—	—	5 T	Gas 8V/3 Auto	All HazMat PPE, one-hour SCBA, Decon, Library, Recovery kits, Spill kits

Figure 7-9 *Apparatus quick reference charts are helpful in evaluating apparatus safety concerns.*

Specialized Equipment Often, the ingenuity and resourcefulness of our firefighters can make a situation more dangerous by using equipment for a purpose other than it was intended. The FDISO should carefully consider this creativity and determine if the makeshift application creates more risk than the benefit achieved from the adaptation. In some cases, specialized equipment may be available. Have you ever seen a traditional engine and truck municipal fire department fight a raging brush fire with firefighters dragging booster lines all over the fire line while a whole fleet of mobile brush patrol vehicles was available from a neighboring department? The FDISO can make a judgment here if, and only if, he or she is aware of the specialized apparatus resources available in the area.

Personal Protective Equipment (PPE)

The FDISO can lose effectiveness if he or she acts as the "Bunker Cop." In the context of this chapter, a bigger picture approach is suggested. The FDISO needs to evaluate if the level of PPE is appropriate to the scope of the incident and the action plan. The FDISO can, and should, make judgment on increasing or decreasing the level of PPE required for operating crews. The classic example is the firefighter dressed in full structural gear when fighting a brush fire. Should the firefighter doff the coat and hood? The FDISO can evaluate fire spread, number of firefighters present, type of attack chosen, as well as the availability of rehab and other resources to make a judgment and advise the Incident Commander. In the case of Hazardous Materials Incidents, the FDISO can meet with the **HazMat** team leader and help ascertain the required levels of PPE for hot zone, warm zone, decontamination, and medical operations. The key to choosing the right specialized equipment is uniquely tied to the potential damage that a process can inflict on personnel. The following are some common hazards that can insult or injure our personnel.

HazMat
shortened word for hazardous materials

Noise Although most firefighters realize that long-term exposure to sirens and air horns will cause damage, few understand that hearing damage is a result of exposure to damaging frequency or loudness, as well as the duration of each of these exposures. Frequency is the length of a sound wave over time. These sound waves cause a vibration of the ear drum and sound may be perceived. Frequency combined with the loudness or decibel (dBA) level of the sound may cause damage. Typically, ultrahigh or extremely low frequencies cause hearing damage at lower decibels and lower duration than a louder, midrange frequency over a longer duration. To show the relationship of duration, consider the following: OSHA allows a maximum permissible noise level of 90 dBA (power saws, gas fans) for 8 hours but a 115 dBA level (air horns, gas escaping under high pressure) for no more than 15 minutes.[4]

■ Note
OSHA allows a maximum permissible noise level of 90 dBA (power saws, gas fans) for 8 hours but a 115 dBA level (air horns, gas escaping under high pressure) for no more than 15 minutes.

Infection Much has been written about the protection of emergency responders from bloodborne and airborne pathogens like hepatitis, the human immunodeficiency virus, tuberculosis, and meningitis. Agencies that routinely provide basic life support (BLS) and advanced life support (ALS) service are mandated to address the whole infection control issue. The FDISO, however, should ensure that firefighting agencies that typically do not provide emergency medical services (EMS) are aware and are provided universal precautions for the times that they are inadvertently placed in EMS roles (mass casualty, firefighter injured, disaster recovery).

Thermal Stress Chapter 4 discussed the physiological effects of thermal stress (hot and cold) on firefighters. One of the few accommodation tools available to fight thermal stress is firefighter PPE. The FDISO needs to monitor the appropriateness of the chosen PPE. Of particular concern, the FDISO should gauge whether the core temperatures of working firefighters will rise to the point of transient heat fatigue or heat stress. In these cases, the FDISO should consider the situation "IDLH" and take steps to accelerate rehab.

Special Operations Specialized operations require unique or nontraditional firefighter protective equipment. Anytime a structural firefighter finds him or herself in a support role of specialized teams, the FDISO should verify that proper PPE is available for that firefighter. Specialized team leaders are a wonderful resource for helping determine this need. Sometimes, it takes an FDISO to bring out these points. Team leaders may feel that the use of a particular piece of PPE is obvious to them, but it may not be obvious to those assisting. The FDISO can view his or her role as the bridge to these situations. Some of these unique PPE items are listed here.

Body harnesses	Fall arresters
Escape air packs	Intercom systems
Splash gear	Core temperature sensors
Portable EKG monitors	Infrared detectors or night goggles
Personal flotation devices	Gas monitors
Ambient heat sensors	Motion detectors
Fire shelters	

One other point regarding the evaluation of PPE. Firefighters are often anxious to doff their gear, which may create a problem. The FDISO should evaluate the necessity of PPE, and give guidance on when equipment can be removed. This approach shows concern for the needs of the firefighters and gives credibility to your task of evaluating risks. Usually, firefighters do not have to be told twice that they can take their gear off. Just be careful that you are not premature with this allowance. One Texas FDISO said that it is routine for him to established a "no bunkers required" zone at all fires. He went on to relay that this technique implied that all the other areas required PPE and that it would be enforced.

BACK TO THE INCIDENT SAFETY OFFICER ACTION MODEL

Resource evaluation is but one component of the Incident Safety Officer Action Model. A careful evaluation of time, personnel, and equipment can help the FDISO make useful recommendations to the Incident Commander. However, remember that the one-time evaluation of these components is not adequate. To be successful, these evaluations need to be repeated cyclically. Combine these evaluations with an incident reconnaissance and careful risk assessment and the FDISO is well on his or her way to *making a difference.*

Summary

One component of the FDISO Action Model is resources. In this arena, the FDISO must evaluate and find hazards relative to time, personnel, and equipment. Time is often forgotten as a resource. The FDISO must avoid misperception of time and should employ some strategies to keep track of time, such as assistance from a dispatch center. Having enough personnel available to handle a given incident is also important. Too few people can lead to an injurious situation just as having too many. The FDISO should also track crews that are committed to risk-exposed situations like IDLH environments. Accountability systems need to be evaluate to see if they are tracking personnel and assignments. The FDISO can check to see that the right equipment is being used appropriately. Failure to do so may lead to injury. Likewise, protective equipment—including specialized protective equipment for noise, infectious disease, and other special concerns—needs to be utilized as appropriate for the situation.

Review Questions

1. What are three resources that need to be evaluated by the FDISO?

2. Describe the relevance of time on an incident and list methods to achieve more accurate timekeeping.

3. From a FDISO perspective, how many firefighters and support personnel are needed to handle a two-room and contents fire in a two-story residential structure?

4. What is meant by "risk-exposed" personnel?

5. List various accountability systems as well as advantages and disadvantages of each.

6. Discuss equipment as a resource.

7. List three common incident hazards and define a method of protection from each.

Notes

1. S. Jerauld, Post Incident Analysis "Builders Square Fire," Palm Beach Fire & Rescue, December 1993.

2. NFPA Investigative Report, *Three Firefighter Fatalities at a Single Family Residential Fire, Pittsburgh, PA* (Quincy, MA: National Fire Protection Association, 1996).

3. NFPA 1500, Fire Department Occupational Safety and Health (Quincy, MA: NFPA, 1992).

4. *Code of Federal Regulation,* Title 29, 1993. 1910.95.

Chapter

8

Reconnaissance— Evaluating the Environment

Learning Objectives

Upon completion of this chapter, you should be able to:

- Discuss the difference between environmental and operational reconnaissance.
- List the three primary areas of environmental reconnaissance.
- Explain the importance of defining a principal hazard during incidents.
- Explain the relevance of reading smoke to predict fire behavior.
- List common principal incident hazards and discuss the elements of each.
- Define four ways to classify the integrity of an incident.
- List common electrical equipment and their associated hazards.
- Discuss weather as a surrounding element and forecast weather's impact on the incident.

One child has already been rescued from the glacial-fed waters. Two adults are still missing after their fishing boat capsized and sank. One adult attempted to swim to shore but is now missing. The other was with the boat when it sank. Six divers are working two separate search patterns. The surface support team has already accomplished witness interviews and is relaying landmarks to the dive team leader. The Incident Commander special-called you, the on-call safety officer, due to the dropping temperatures and the outlook of a potentially long and emotional rescue and recovery effort for your department's dive team and surface support people.

As you arrive and get briefed, you immediately realize that the environment itself will pose significant hazards. The rocky lake shore, limited access, frigid water, and dropping temperatures have you concerned. So far, the teams are task focused and doing well. You know however, that the switch from rescue to recovery will heighten the chance for injury just in terms of environmental concerns. The Incident Commander asks you to make a good reconnaissance observation and bring back to him the hazards you think should be addressed first. Specifically, you know you should start looking at. . . .

Another area in the Incident Safety Officer Action Model is that of reconnaissance or recon (Figure 8-1). Often, recon is the first assignment given the Incident Safety Officer by the Incident Commander (IC), for simple reasons. The IC needs a thorough 360° size-up of the incident scene, hazards, and operations. How is the

THE FIRE DEPARTMENT INCIDENT SAFETY OFFICER ACTION MODEL

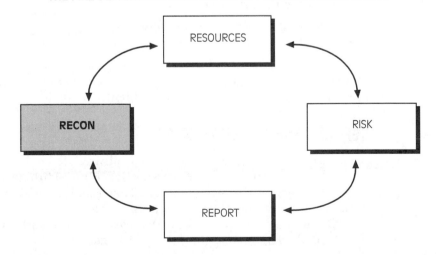

Figure 8-1 *The "recon" box of the FDISO Action Model reminds the FDISO to make multiple operational and environmental reconnaissance trips.*

FDISO size-up different than the standard initial size-up we learned about in basic fire school? The difference is found in the premise that the FDISO should have the skill to articulate an accurate, descriptive, and interpreted survey of the incident environment and operation; whereas the basic size-up we learned of originally is a mere mental process of factors and probabilities of the incident.

The reconnaissance effort should come early in the incident and be repeated as conditions change or on a timely basis. It is not uncommon for an FDISO to perform eight, ten, even twenty recon trips during a working structure fire. Wildland fire incidents may require several dozen recon trips where the FDISO is flown over the incident and is accompanied by a Fire Behaviorist, Plans Section Chief, or maybe even the Incident Commander. The recon trip is an opportunity for the FDISO to look at many conditions and details that spell out the big picture for defining *risk*. Often, the recon trip is merely watching the environment and operations underway. While watching, the FDISO considers the nuts and bolts we discussed in Chapter 4. Experience, knowledge, theory, study, and wisdom combined with the visual image collected with the recon effort allow the FDISO to make a determination, judgment, or advisement on the potential or probability for firefighter injury. In other words, the FDISO is *evaluating* the incident.

When making an incident reconnaissance, the FDISO can divide observations into two categories: environmental and operational. Although each incident presents unique environmental and operational concerns, this chapter covers basic concerns in regard to evaluating the environment; Chapter 9 covers evaluating operations.

For the context of this book, consider the environment to be the many features, details, and conditions that are physically present. Also include any potential changes to the physical makeup of the scene. Although firefighting efforts change the environment, we discuss those impacts under Evaluating the Operation in Chapter 9. The environment can be evaluated by considering three steps:

1. Defining the *principal hazard*
2. Defining the *environmental integrity*
3. Defining the *effects of surrounding elements*

In each of these, the FDISO is challenged to not only evaluate what the current condition of each is, but also to predict or anticipate the change to each of these and how each will affect the incident.

DEFINING THE PRINCIPAL HAZARD

If a free-burning fire is traveling down a hallway of a center hall apartment building, the hazard, from the FDISO perspective, is rapid fire spread (Figure 8-2). This may seem obvious. However, the FDISO needs to look at the big picture and determine what the principal hazard is to firefighters.

Often, the primary hazard is the reason the fire and rescue department has been called. Sometimes, the principal hazard is a loss of integrity (like in a

■ Note
It is not uncommon for an FDISO to perform eight, ten, or even twenty recon trips during a working structure fire.

■ Note
Experience, knowledge, theory, study, and wisdom combined with the visual image collected with the recon effort allow the FDISO to make a determination, judgment, or advisement on the potential or probability for firefighter injury.

Figure 8-2 *A fire that has captured a central hallway will lead to rapid fire spread, the principal hazard at this incident.*

building collapse or trench rescue) or maybe the principal hazard is fast-moving water. At times, it is not so much that a fire is burning, it is the location of the fire that has created the principal hazard. The previous example of a center hallway fire in an apartment building illustrates this. At times, the principal hazard is less obvious. Usually this is the case with motor vehicle accidents—the FDISO has to find the principal hazard. It could be fuel containment, vehicle instability, other traffic, or access to the scene. If a circus tent blows over, the principal hazard may be secondary collapse potential. In most cases, the principal hazard and its location provide a foundation for the FDISO to predict what will happen next. Following is a list of three principal hazards, interior fires, wildland fires, and collapses, with more specific descriptions of the principal hazard with regard to predicting changes.

Interior Fires

Phase or Stage of Fire Observations of visible smoke (fire gases) and fire to determine the size of the fire, the speed of spread, the degree of heat release, and the potential for flashover, smoke explosion, and/or back draft are important at this stage. These observations are often made from the outside of the structure. If the fire is visible, its behavior is a bit more obvious. More often than not, the smoke visible outside the structure must be evaluated to determine the phase or stage of the fire.

Figure 8-3 *Smoke density indicates the quality of the burning process.*

This is the art of reading smoke. More specifically, we need to look at smoke characteristics and what each characteristic may indicate.

Smoke density usually indicates the quality of the burning process (Figure 8-3). High-density smoke usually means incomplete burning of fuels due to low heat or insufficient combustion air. Light or thin smoke may indicate complete burning, good ventilation (dissipation), or incipient burning. Moderate smoke density that remains constant indicates a free-burning fire with thermal balance. Smoke that rapidly becomes dense indicates an imminent flashover. Density that rapidly changes from heavy to light may indicate that a flashover just took place (you should see a change in velocity) or that the fire has been adequately vented.

Smoke velocity (Figure 8-4) usually indicates the rate of heat release and the speed of spread of the fire. Slow-moving smoke is indicative of slow fire spread or cooling of smoke. Fast-moving smoke may indicate rapid fire spread or extreme heat. Smoke that leaves the building fast but then slows is an example of a smoke-filled box with little room left for expansion. Such a building is a candidate for flashover or smoke explosion. Intermittent smoke release is indicative of a backdraft potential or a situation where external winds are "opening and closing" the vent port. "Boiling" or "chugging" smoke is a condition where high-velocity, high-density smoke fills a ventilation opening (door, window, or hole) and is expanding faster than the "box" can release the accompanying pressure. This situation indicates a well-involved, extremely hot fire that is capable of a rapid, room-to-room flashover and full involvement of the structure. Autoignition of these

Figure 8-4   *Smoke velocity indicates the rate of heat release.*

released gases may take place just outside of the vent opening, prior to the flashover. Also worth considering here is the relationship of the outside temperature. A hot, dry day may slow down the smoke whereas cold, dry air makes smoke velocity appear more rapid. Moisture in air also slows convected currents.

Smoke volume most often indicates the amount of fuels burning (Figure 8-5). A large quantity of smoke is indicative of a heavy fire fuel load or the buildup of smoke prior to release. A small volume of smoke indicates a small fire or complete burning (fire gases may not be visible). Once again, a rapid change in apparent smoke volume may indicate fire events. Volume that rapidly diminishes may indicate that a flashover just occurred (burning off a volume of smoke and fire gases) or that the gases within a box have been released and the fire is slowly advancing; the velocity of the smoke will differentiate the two. Volume that rapidly increases means that a heavy fuel load has just been added to an already hot fire. Additionally, a door that was containing fire may have just failed, causing fire spread and release of pent up gaseous fuels.

Smoke color most often indicates the types of fuels burning (Figure 8-6). Our previous fire behavior and chemistry classes taught us that petroleum products release black plumes whereas forest and field vegetation emit white and grey plumes. A large quantity of white and grey smoke seen rising in the distance is known as a "**loom-up.**" This wildland firefighting term is used to describe a hot, advancing wildland fire in trees, brush, or grasses. Typically, a visible loom-up

loom-up
a large quantity of smoke that can be seen rising from a distance, usually indicative of a working, hostile fire

Figure 8-5 *Smoke volume is indicative of the quantity of involved fuels. This "loom-up" indicates a significant quantity of burning petroleum products.*

is an indicator of a serious fire requiring additional resources. A uniform color of smoke can help you classify the primary fuel involved, therefore helping you understand the fire behavior expected. In most structural fires, however, the smoke shows various colors and hues that appropriately indicate the many types of fuels that may be burning. High-density, high-volume smoke will almost always be dark grey with hues of black. Most interior fires release similarly colored smoke. Any unusual color (green, red, orange, etc.) is indicative of a burning hazardous material. The release of a small quantity of lazy, dense, and yellowish smoke may indicate a back draft potential. This release usually occurs when small, concealed building voids are ready to back draft. A uniform, expanding white gas is most likely steam from extinguishment efforts.

Location of Fire Fires should be classified as either room (or rooms) and contents fires or structural fires, the latter meaning the burning or exposure of the

Safety
When the fire has captured the structural components of the building, immediate attention to collapse potential must be paid by all firefighters on scene.

■ Note
The successful incident commander looks at the location of the fire and predicts where the fire will move to next, and so should the FDISO.

Safety
Fires in void spaces and above drop ceilings pose additional threats to working groups. Pay particular attention to these construction features—firefighter safety depends on it.

building's structural, load-bearing components. When the fire has captured the structural components of the building, immediate attention to collapse potential must be paid by all firefighters on scene. Fires on lower floors expose upper floors. Upper floor fires create access problems. Fires that have captured a stairwell or vertical chute spread rapidly. Roof fires can cause fires in the ceiling space below the roof, especially if the roof is a built-up material on a metal deck. Fires in utility and air distribution systems may also distribute the fire to remote locations. Fires in small rooms develop and spread quickly to other rooms but with little force. Fires in large rooms develop slowly but "blow torch" through halls and other smaller rooms or openings. The successful incident commander looks at the location of the fire and predicts where the fire will move to next and so should the FDISO. However, the FDISO needs to predict that movement and communicate it to the IC so that crews can be positioned appropriately. A reminder here: Fires in void spaces and above drop ceilings pose additional

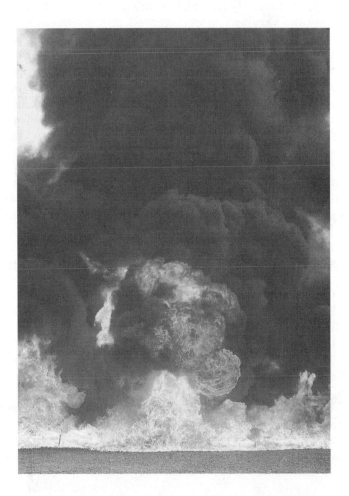

Figure 8-6 *Smoke color can help the FDISO determine a general fuel classification that is contributing to the fire.*

threats to working groups. Pay particular attention to these construction features—firefighter safety depends on it.

Wildland Fires

Location of Fire Most wildland firefighting texts refer to the location of the fire as the "topography." **Topography** refers to the lay of the land in respect to elevation, slope, and aspect (relation to the sun). For the FDISO, the location of the fire needs to be evaluated with regard to spread potential and access for firefighters. All other factors remaining constant, fires burning on the sun-exposed side spread faster than those on the shade side. Likewise, fires burn faster uphill than downhill. Canyons, draws, chutes, and chimneys tend to "shape" the wind and therefore the fire. Often, these geographical features cause wind eddies that can create fire ribbons or whirls (like a dust devil) and other dangerous fire phenomena. Elevation affects humidity (the higher, the dryer), thermal layers (leading to smoke inversions and varying fire spread rates), and wind, each of which can affect fire spread. In the case of an I-Zone fire requiring structural protection, the FDISO needs to evaluate the location of the structure as it relates to **defensible space.** At a minimum, a 30-foot clear-fuel space needs to exist between the structure and the approaching fire. A wind-driven canopy fire may defy this 30-foot clear zone and still threaten the structure. Time and wind are also considered when evaluating the degree of risk facing protection crews. The FDISO can, and should, form a risk opinion based on solid understanding of the location of the fire and its relationship to working crews.

Fuels Involved During a wildland fire, the FDISO can evaluate fuels and determine if a given attack method or exposure protection assignment will be effective. Fuels are considered in three ways: The type, continuity, and moisture content of the involved or potentially involved fuels should each be evaluated.

Type of Fuel Surface fuels like grass, brush, slash, and litter typically burn fast and hot due to the amount of exposed surface area. The oil content of some of these fuels significantly increases the heat release level of the fire. Dry yucca, scrub oak, and other high-oil content fuels can cause explosive fire spread. Fires in tree tops (aerial, crown, or canopy fire) are extremely dangerous and often mean that a direct fire attack with ground crews should be avoided.

Continuity of Fuels The continuity of fuels can be evaluated with regard to arrangement and height. Evaluating the arrangement of fuels can help determine if the direction and speed of fire spread, given the prevailing wind and slope, are a danger to personnel. The height of fuels can precipitate "ladder fuels." Ladder fuels are responsible for elevating a surface fire to a canopy fire by climbing intermediate fuels. This continuity directly affects crew safety and the FDISO needs to understand this relationship.

topography
accurate description of a place or the lay of the land

defensible space
clear fuel space around a structure; a minimum of 30 feet of cleared area that allows for the defense of a structure during wildland and I-zone fires

Moisture Content The FDISO needs to pay attention to fuel content reports as they indicate the fire spread potential. For the sake of classifying fuels, dead vegetation is used and broken into time units. For example, a 10-hour fuel is one that the moisture contained matches the relative humidity of the surrounding air in 10 hours. One-hour fuels include grasses, light brush, and twigs less than 1/4 inch in diameter. Ten-hour fuels included most brush, limbs less than 1 inch in diameter, and pinecones. One hundred-hour fuels and one thousand-hour fuels are the larger fallen logs and standing timber. Moisture content below 25% for 1-hour fuels and below 15% for 10-hour fuels should be cause for alarm.

Weather   Weather was discussed in Chapter 4 and is discussed more in depth later in this chapter under Weather. Key here is the relationship of humidity, wind, and potential for change, each of which can create a hazard where none previously existed. The rapid arrival of a cold front created a 180° wind shift that caused the blowup that killed fourteen firefighters at the Storm King Fire in Colorado.

Collapses (No Fire)

Location of Collapse   In many collapse scenarios, the principal hazard is the potential for a secondary collapse. This concern is covered shortly in the defining of environmental integrity. However, some considerations need to be covered in defining collapse hazards. The location of the collapse can indicate contributing factors toward a secondary collapse. A collapse near freeways, railyards, or airports is susceptible to vibrations of the surrounding earth due to the presence of the respective traffic. Collapse of earth or structures in and around utility distribution systems is cause for careful examination. Breached water and sewer systems should cause immediate alarm for secondary collapse and are difficult to abate. Turning off a water pipe does not necessarily remove the hazard. Gas and electrical systems are easier to control but add an additional fire and life safety hazard when not controlled.

Material Involved   To further define the principal hazard in a collapse, the FDISO needs to evaluate the material involved. A building collapse contains primarily steel, wood, or masonry as does a bridge collapse. A landslide or trench collapse primarily involves earth and vegetation. Each of these has characteristics that help predict the potential for secondary collapse. When evaluating structural collapse, it is important to consider that the imposition of loads has undoubtedly changed. What once was a column (compressive load) may now be a beam (tension and compression). If a column has twisted, a torsional load may now be present. This new load imposition should be compared to the material of the member. Steel is very resistive to compression and tension, assuming the load is axial. A compressive, torsional load can cause buckling and twisting of steel unless mass is present. Masonry, on the other hand, is weak to tension and to torsion, but extremely resistive to compression. Wood is relatively poor in all categories. The shape of the material should also be considered. If a 12-inch high I beam is turned on its side, it no longer has the strength of its height. It probably has the strength of a 3-inch high beam. If a load is still present on the beam, the beam is on the verge of a deflection that could drop the remaining load.

Earthen collapse can be equally complicated. The FDISO can sort this out by evaluating the water content, density, and type of earth involved. Water content affects weight as well as lubrication of the material, which are both directly proportional to collapse potential. The higher the water content, the higher the potential for collapse. The density of an earthen material can affect the ability of the material to repel or absorb water. Sand, clay, topsoil, gravel, and talus each has its own characteristics. Sand collapses easily but is very predictable; clay is more resistive to collapse but less predictable. Gravel will collapse into a classic "pyramid" slope. Talus is actually a coarse gravel and is therefore more resistant to movement compared to gravel.

Load Distribution The FDISO needs to visualize the new collapse arrangement and trace loads. This visualization may seem to be a science but can equally be an art. Using a strong understanding of the imposition of loads, the FDISO can look at the initial collapse and "see" where the redistribution of dead and live loads exist. Some warnings should be present, especially with modern, lightweight buildings. These warnings are typically found by seeing a beam that is now a column. Likewise, what once was a fixed beam may now be a cantilevered beam. One effective method for finding incongruities is to step back and find the greatest load. The load may be a roof assembly, mechanical device, building content, or a collection of many loads. Once the load is determined, picture how it is being distributed along building components to the ground. As you visually trace this path, determine if loads are being compressed through a material, twisting a material, or pulling on a material. In some cases, the load may be trying to shear a material. Each of these situations may present as inappropriate for the material involved. By visualizing loads and finding incongruities with the materials carrying the load, the FDISO can predict secondary collapse and therefore evaluate the risk of crews working in these areas or the steps necessary to reduce exposure.

Typically, dirt and rocks settle to a classic pyramidal shape. Obstacles can hamper this settling. If an obstacle prevents the earth from settling in a pyramid fashion, then the obstacle needs to be evaluated for whether it can handle the new load. Following a landslide or trench collapse, the remaining walls (or face) will be free of vegetation or other bracing elements, leading to the potential for a secondary slide. Like a good bomb technician, the FDISO is always looking for the second bomb, or in this case, the next slide.

As you can see from the foregoing information, a primary step in evaluating the environment is to identify the principal hazard and predict what will happen next. Once the principal hazard is defined, the FDISO can evaluate the overall integrity of the incident.

■ **Note**

By visualizing loads and finding incongruities with the materials carrying the load, the FDISO can predict secondary collapse and therefore evaluate the risk of crews working in these areas or the steps necessary to reduce exposure.

DEFINING ENVIRONMENTAL INTEGRITY

Environmental integrity is the state of being sound, whole, or intact. The FDISO is responsible for determining the quality of this state. The environment we, as firefighters, encounter is comprised of buildings, vehicles, access ways, weather, physical features, topographical features, and people. To evaluate the integrity of each of these seems impractical. Each, however, may affect the safety of our operations. Additionally, the FDISO must determine the degree that each environmental consideration affects the overall incident. Usually we do this by categorizing the integrity consideration as stable or unstable. A better way to categorize is as follows:

> Stable—Not likely to change
>
> Stable—Might change
>
> Unstable—Changing
>
> Unstable—Rapidly changing

So which environmental considerations require evaluation and how do you determine integrity? Here, we list important considerations regarding integrity of buildings, utilities, equipment, and surroundings and give some specific information to help evaluate each of these to determine integrity.

Building Integrity

The FDISO *must* determine the structural stability of a building being attacked by fire; incident commanders expect this of the FDISO. Building construction knowledge, fire behavior understanding, and firefighting experience combine here to help the FDISO make this determination. This chapter cannot replace the information found in Chapter 4 or in recognized fire service building construction texts. The information listed here, however, is provided to assist the FDISO in framing a decision tree for evaluating building integrity.

Size and Type of Building General classifications can help the FDISO determine the susceptibility of the building to collapse (Figure 8-7). A single-family dwelling roof will typically collapse in small segments due to the number of partition walls supporting rafters or trusses. Garages and great rooms are exceptions to this behavior. Unfinished basements can lead to a catastrophic collapse of the main floor due to the exposure of the main girder or floor beams. The occupancy type of a commercial structure needs to be determined prior to collapse prediction. A hospital may be built to resist significant collapse whereas a medical plaza or doctor's office may be built no differently than a neighborhood convenience store, or, in some jurisdictions, a single-family dwelling. Any process or production that takes place within a structure can help determine collapse potential. Storage structures typically collapse early due to weight and configuration of the stored

Figure 8-7 *The size and type of building directly affect its integrity during a fire.*

material. Collapse of the stored material, independent of the building, is also a concern. Processes involving flammables can accelerate the fire spread and heat levels leading to rapid structural failure, especially in buildings with exposed steel structural members. Extremely large buildings have lower mass versus size ratios leading to earlier failure of a given portion of the structure.

Construction Features Certain construction features should be viewed as a red flag for collapse (Figure 8-8). Trusses of virtually any type need to be included in this category. Multiple roof lines, additions, remodels, and construction in progress suggest early collapse under fire conditions. Businesses, occupancies, or buildings that share common ceilings, basements, and utilities are candidates for early collapse. The presence of falsework is an excellent clue that collapse will be early and significant. Falsework is temporary shoring used to support floors, roofs, and walls during construction or during concrete pouring. Heat, ventilation, and air conditioning (HVAC) systems should be evaluated to determine if they increase the collapse potential of the structure. Rooftop HVAC units are often a concentrated load visible on top of the roof. What may not be visible is the array of ductwork, hoods, and other components that are hung under the roof. Signs, decorations, marquees, balconies, facades, and cornices are often nailed or hung on the exterior of the building; these can fall off early in the fire.

Figure 8-8
Construction features such as major renovation and additions affect building integrity.

Building Access Access to and maneuverability around a building are equal components in evaluating building integrity. Is the building accessible to all responders? What risk do the responders encounter by parking near the building? Is traffic an inherited problem while operating near the building? What obstacles does the building present to crews maneuvering around? Are fences, gardens, gates, terracing, trees, ledges, and other parked vehicles presenting access problems? A mass evacuation may create a substantial obstacle. Do overhangs, overhead power feeds, narrow alleyways, or other close buildings present an obstacle? Are all doors and windows available for firefighter access and rapid egress? The integrity of the building also means integrity of its access.

Utility Integrity

The evaluation of utility integrity is essential due to the significant life-safety risk they pose (Figure 8-9). Electricity, gas, water, sewer, and communications equipment need to be evaluated. Previously, fire departments would simply shut down all known utilities to a given structure for the sake of fire fighting. This action has led to premature shutdown of essential fire suppression and protection features within the structure, leading to further damage. Fire and rescue departments now evaluate which, if any utilities to shut down, therefore an evaluation of utility integrity becomes important.

Figure 8-9 *The FDISO should determine utility integrity during recon.*

Electrical Systems The integrity of electrical systems is based on the system being properly grounded, insulated, and circuit protected. A disruption of any of these components can create danger to the firefighting operation. Figures 8-10 and 8-11 show common electrical language and hazards.

Fire and rescue departments are often called to any incident that involves electrical distribution equipment (cat on the pole, wires down, etc.). Rarely, however, do the firefighters become involved in solving the incident without the assistance from the local power company. Occasionally, though, the firefighters must act without the power company to save lives or stop a potential loss. In these cases, a basic understanding of electric system integrity is important. For the Incident Safety Officer, understanding these systems is *essential*. At structure fires and accidents, the FDISO should evaluate the integrity of electrical systems. When a component of the total system is deemed to have lost integrity, an electrical danger exists and needs to be communicated to personnel working the incident. In some cases, the electrical system can be perfectly intact and still pose a threat to firefighters. This is true in cases where aerial apparatus is working near power lines. Certain electrical equipment boxes also pose threats to firefighters (Figure 8-11).

Gas Utilities Firefighters who have actually witnessed a propane tank BLEVE (Boiling Liquid Expanding Vapor Explosion) will undoubtedly argue that the most

!Safety

At structure fires and accidents, the FDISO should evaluate the integrity of electrical systems. When a component of the system is deemed to have lost integrity, an electrical danger exists and needs to be communicated to personnel working the incident.

important utility to control is gas. The FDISO should make an effort to evaluate the integrity of the gas supply and containment vessels. The integrity of a gas system relies on a tight supply vessel (tank or piping), shutoff valve, pressure regulation device, and distribution system with protection at each appliance (shutoff valve and surge protection). Each component in the system is designed for a certain pressure. Pressures that exceed this amount (like exposure from a hostile fire) can cause a pressure relief device to activate, introducing expanding gas into the

---

Electrical Terms

Ampere (Amps)
The unit of measure for VOLUME of current flow.

Continuity
The completeness of a current circuit; the ability of electricity to pass unbroken from one point to the next.

Conductivity
The measure of the ability of a material to pass electrical current.

Energized
The presence of electrical current within a material or component.

Ground
The position or portion of an electrical circuit that is at zero potential with respect to earth. The point of electrical return for an electrical circuit.

Ground Fault Interruption (GFI)
A device that will break continuity in a circuit when grounding occurs before the current returns to the distribution source.

Grounding
The act or event of creating a point for electrical current to return to zero potential.

Ohm
The unit of measure for resistance of electrical current. One ohm equals the resistance of 1 volt across a terminal at 1 ampere.

Resistance
The degree that a material holds or impedes the flow of electrical current.

Static Discharge
The release of electrical energy that has accumulated on an insulated body. Static is a stationary charge looking for a ground.

Voltage/Volt
The FORCE that causes the flow of electricity. A volt is the unit of measure for electrical force or potential.

Watts
A unit of measure for the amount of energy a specific appliance uses.

Figure 8-10 *The FDISO should be comfortable with electrical terms and be able to communicate hazards accurately.*

Electrical Equipment Hazards

Powerlines/Wires
Uninsulated, under tension, arc danger, difficult to know voltage/amperage, downed wires may jump/recoil, ground/fences/gutters easily energized when in contact, power feed may be from both directions.

Pole-Mounted Transformers
Usually step-down type, difficult to extinguish, may drop/dangle, may cause pole damage, may cause wire failure, may drip hot oil and start ground fire.

Pad-Mounted Transformers
Usually low-voltage/high-amperage, may energize surrounding surfaces, pooled water may conduct current, difficult to extinguish oil/pitch fire, possibility of arcing.

Ground Level Vaults
Confined space, possible O_2 deficiency, buildup of explosive gases/smoke, cable tunnels can transfer fire/heat, significant arc danger.

Subterranean Vaults
Same as ground level vaults, plus water collection hazard, difficult to ventilate, can "launch" a manhole cover if accumulated gases ignite.

Generators
Power source, tremendous heat generation, hazards of diesel/gasoline/gas-fueled engines, automatic startup when other power sources are disconnected.

Batteries
Stored energy, chemical/spill hazards, explosive gas buildup, multiple exposed terminals.

Disconnects/Switches/Meters
Exposed terminals, danger of arcing.

Figure 8-11 *Common electrical components and their associated hazards to firefighters.*

environment. Likewise, trauma to this system can create holes, pipe separation, or container failure. Once gas escapes from the system, fires may be accelerated or toxins can be inhaled. Unignited gases accumulate in confined spaces and present an explosion hazard if an ignition source and proper air mixtures are introduced. Knowing the properties of common gases can help the FDISO determine the risks associated with the gas present (Figure 8-12).

Water and Sewer Water and sewer systems can cause safety concern in collapse situations. Uncontrolled water flow can cause initial and secondary collapse in structures. A damaged sewer or storm drain system can leech into surrounding gravel and dirt beds, undermining the very earth that may support a structure, road, or other utility system. Firefighting efforts may introduce significant quantities of water into a structure. When this happens, who evaluates where all that water is going to go? Likewise, the observation of large quantities of water flowing in and around the incident should warrant an investigation by the FDISO. In

Properties of Common Utility Gases		
Property	**Propane (Liquefied Petroleum Gas)**	**Natural Gas (Methane)**
Chemical makeup	C_3H_8	CH_4
Vapor density	1.5	.55
Boiling point	−44°F	−259°F
Ignition temperature	871°F	999°F
Flammability limits		
Upper	9.5	14.0
Lower	2.4	5.3

Figure 8-12 *The FDISO who knows gas properties can more effectively assess risks associated with utilities.*

one Loveland, Colorado, incident, the FDISO saw large quantities of water pooling in the yard next to a dwelling fire. This seemed unusual to the FDISO. Upon investigation, the FDISO determined that piping feeding the hydrant being used for water supply had begun leaking underground. The pooling was the first visual clue that the ground underneath the firefighting operation was being undermined. The water department was notified and the water shut down but not before a 8-foot deep sinkhole collapsed an adjoining roadway 20 feet from the first-due engine!

Vehicle Integrity

For the sake of this text, all sorts of transportation equipment, such as cars, trucks, boats, planes, and trains, can be classified as vehicles (Figure 8-13). When evaluating vehicle integrity, many of the same concepts presented earlier in this chapter apply. Specifically, though, consider the following:

Stability and Position Is a vehicle resting on all its wheels stable? Most extrication experts say absolutely not! The simple act of deflating tires or chocking wheels is standard practice for many rescue companies. The FDISO can evaluate the position of the vehicle as it relates to the slope of the road or potential to roll. If the vehicle were to roll, where would it go and who or what would get dragged along or struck? One question that an FDISO could ask him- or herself would include, "If another car were to strike this one, what would the result be?" Often, the thought of such an event will cause the FDISO to put in place a higher degree of barricading or scene screening so that other traffic does not become part of the incident (Figure 8-14).

Vehicles that are not on their wheels can create additional safety concerns for rescue crews. The FDISO obviously should evaluate the efforts and tech-

Figure 8-13 *Vehicle incidents present their own spectrum of integrity issues.*

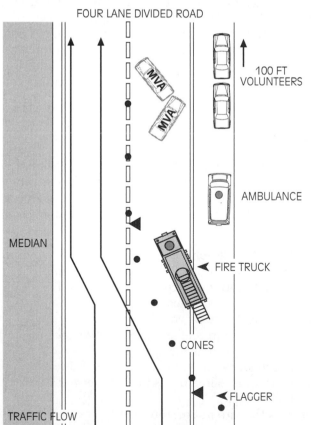

Figure 8-14 *The FDISO should evaluate the relationship of moving traffic to the incident. Are crews protected?*

niques employed to stabilize the vehicle. Using load imposition knowledge, determine if the stabilization technique is adequate. Likewise, evaluate the surface that the vehicle is resting upon. Will the surface support the load? Will the addition of water, fuel, oil, rescuers, or tools create an unsafe environment? Many experienced firefighters can recall a situation when they were extricating a victim only to have the vehicle prematurely move. The effective FDISO predicts this situation and communicates accordingly (Figure 8-15).

Status of Vehicle Systems At a structure fire, a crew is usually assigned to shut down or isolate utilities. A vehicle incident should be treated similarly. If a crew has not been assigned this task, the FDISO should evaluate the vehicle systems to see if they will contribute to the instability or risk of the incident. These systems include the following as a minimum.

Fuel System The fuel system is comprised usually of one or more reservoirs, a pump, conveyance or supply lines, and some kind of distribution system (carburetor, regulator, injectors, etc.). Each of these should be checked for leaks or separation. Most fuel systems are under pressure so leaks are obvious. Alternative fuel systems such as compressed natural gas, propane, and stored electricity have similar makeup. On some cars, the ignition switch also activates the fuel pump or opens fuel delivery valves. If the ignition is still on, consideration

Figure 8-15 *Vehicles not on their wheels present crews with multiple hazards relative to scene integrity.*

should be given to turning it off. In the case of leaks, standard protocol for operations in flammable environments are appropriate. Use of aqueous film forming foam (AFFF), absorbent material, or other spill control is warranted. Ignition sources should be isolated and controlled, including the use of extrication tools.

Electrical System The primary concern with the electrical system is the status of the ignition and battery. Many rescue technicians argue about the merits or dangers of disconnecting the battery. With the inclusion of passenger restraint devices such as air bags, the disconnect of the battery may be a positive step for improving the scene safety. Some vehicles (like the Jaguar) do not depend on the battery to help initiate air bag deployment. Accidental deployment of air restraint devices has and can cause injury to rescuers. C-spine trauma, muscle sprain, contusions, and lacerations are possible if the unsuspecting firefighter is suddenly hit by a late-deploying restraint bag. Late in 1996, an infant died from a delayed activation of an inflatable restraint device. If the decision to disconnect the battery is made, common sense dictates that an evaluation of fire potential comes first. To reduce the chance of an arc during disconnection or cutting of wires, see that all electrical devices are turned off in the vehicle and that the negative lead (on most vehicles) is disconnected first. The suggestion is not that the FDISO performs these steps, but rather that the FDISO communicates these hazards and procedures as a safety reminder to the crew assigned the task.

!Safety Accidental deployment of air restraint devices has and can cause injury to rescuers. C-spine trauma, muscle sprain, contusions, and lacerations are possible if the firefighter is hit by a late-deploying air restraint bag.

Access to Vehicle Interior In a recent training exercise, the extrication instructor showed what happens when a simple B-post is cut when attempting rescue of persons trapped in a rolled-over minivan. The students should have been warned of the impending problem when the hydraulic cutting tool became jammed in the cut. Once the cut was made, the weight of the vehicle slowly folded the remaining posts. When evaluating vehicle integrity, the extrication effort to provide access to victims may be the factor causing a degradating integrity.

DEFINING THE EFFECTS OF SURROUNDING ELEMENTS

The final component in the environmental reconnaissance effort is evaluating the effects of surrounding elements. This can be viewed as a catchall category but is equally important as the previous elements. Some miscellaneous elements that need to be evaluated include physical surroundings, access, and weather factors.

Physical Surroundings

As the FDISO makes his or her way around the incident scene, he or she should establish a relationship of any physical exposures that may affect the outcome of

■ **Note**
The effective FDISO evaluates any physical item, including terrain, foliage, curbs, posts, fences, drainages, and barriers and decides if that item could affect the operation.

the incident—notice that we did not say affect the safety of an incident. Too often, we think of the FDISO as only looking out for life safety items like trip and fall hazards, overhead obstructions, and so forth. The effective FDISO evaluates any physical item, including terrain, foliage, curbs, posts, fences, drainages, and barriers and decides if that item could affect the operation.

One physical feature, a sloping grade, has been cited as a significant factor leading to multiple firefighter deaths.[1,2] In slope–grade incidents (see Figure 8-16), a crew that thinks it is working on the first floor may in fact be working on the second floor. Usually, the presence of a fence or other barrier has masked this situation from the IC or the advancing crews.

It is easy to see how communications can be mixed up. The importance of communicating fire location, conditions, and building features cannot be overemphasized. Often, the FDISO is the first firefighter to get a detailed 360° look at the

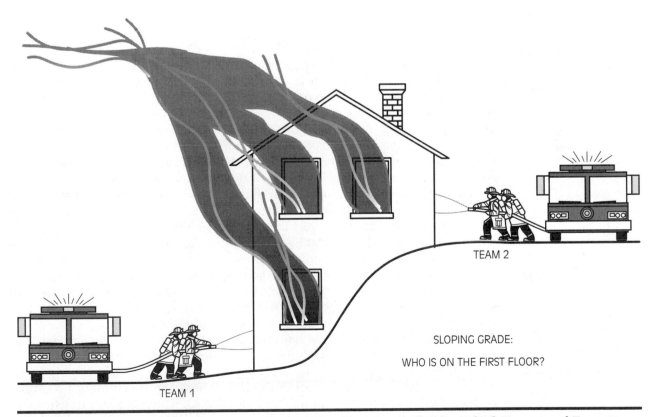

TEAM 2

SLOPING GRADE:

WHO IS ON THE FIRST FLOOR?

TEAM 1

Figure 8-16 *Sloping grades may cause dangerous miscommunication. In this figure both Team 1 and Team 2 may believe they are on the first floor.*

building and thus is obligated to communicate to the IC the presence of a slope grade as well as an observation about the location of the fire in relation to attack crews.

Access

The presence of fences, shrubbery, and other barriers may impede incident operations by restricting access or impairing an adequate size-up. The FDISO can determine the relevance of barriers to the incident.

As a rule, the FDISO should survey the entire area of incident impact and ensure that each crew has at least two escape areas (safe havens) as well as a clear corridor in which to egress. If this is impractical or impossible to achieve, the risk to that crew should be considered increased and the FDISO should include this information in his or her hazard report to the Incident Commander. Additionally, escape routes and any escape barriers need to be communicated to the affected crew.

Weather

You have probably heard the old adage, "If you don't like the weather, just wait an hour and it will change!" It is amazing how many regions have adopted this phrase to capture their "unique" weather considerations. Weather is a dynamic, complex, and often misunderstood force that firefighters must contend with. Common firehouse talk suggests that the "big one" will hit when the weather is at its worst; perhaps this should be another one of Murphy's laws. Often, the adverse weather has caused the reporting party to call for our service. Progressive fire and rescue departments have acknowledged this fact and taken steps to reduce the impact of adverse weather. In Chapter 3, we made a case for the automatic inclusion of an FDISO for incidents that occur when weather conditions reach extremes for that locale. Once on scene, the FDISO can weigh the effects of weather extremes with the behavior of the incident as well as the operations underway.

The effective FDISO studies weather and understands the particulars of weather patterns found in his or her geographical area. Likewise, the effective FDISO keeps abreast of daily forecasts and weather observations as a matter of habit and readiness. From this constant attentiveness, the FDISO can make reasonable predictions on the impacts of weather-related risks.

This particular section is not intended as a replacement for professional weather spotter and forecasting training—these are available from local television weather forecasters and the National Weather Service (often free of charge). The National Weather Service is part of the U.S. Commerce Department's National Oceanic and Atmospheric Administration (NOAA), which has an extensive collection of pamphlets, books, videos, and charts available for weather-related education. This section of the book highlights weather considerations essential to effective FDISO performance. These considerations are wind, humidity, temper-

Safety
The FDISO should survey the entire area of incident impact and ensure that each crew has at least two escape areas (safe havens) as well as a clear corridor in which to egress.

▪ Note
The effective FDISO studies weather and understands the particulars of weather patterns found in his or her geographical area.

ature, and potential for change.

Wind Of all the weather considerations affecting fire fighting operations, wind is by far the most important. Nothing can change a situation faster or cause more frustration in dealing with an operation than the effects of wind.

Wind is created as air masses attempt to reach equilibrium. Warm and cold fronts cause changes in atmospheric pressure and therefore gradients in pressure. Other factors such as the jet stream (prevailing wind), day/night effect (diurnal wind), upslope/downslope, and sea breeze all influence wind. The arrival of a cold front can cause a 180° change in the wind direction. A falling barometer (measure of atmospheric pressure) can indicate an approaching storm and subsequent wind. A foehn wind (pronounced "fain") is a warm, dry wind that spills down through mountain canyons or valleys. A strong foehn wind can be extremely dangerous to firefighters in that it quickly changes fire behavior. Likewise, the sudden reduction of a foehn wind allows the local or prevailing wind to become the influence for the fire. A sudden calm period may indicate an upcoming weather event such as a thunderstorm downdraft or the arrival of a wind shift.

■ **Note**
Of all the weather considerations affecting fire fighting operations, wind is by far the most important.

Some indicators of wind and wind changes can be found in the patterns formed by clouds. High, fast-moving clouds may indicate a coming change, especially if the clouds are moving a different direction than surface wind. Lenticular clouds are very light, sail-shaped cloud formations found high aloft. These indicate high winds that may produce strong downslope winds if they surface. The development and subsequent release of a thunderstorm causes erratic winds and strong downdrafts (Figure 8-17). It is not uncommon to have four or five wind direction and speed changes in the vicinity of a thunderstorm squall line.

It is important for the FDISO to understand the wind patterns for his or her specific geographical region. Further, the FDISO should evaluate existing and forecasted winds and predict their influence on fire behavior and safety of the personnel.

As wind velocity increases, so does the risk to firefighters. Knowing local wind tendencies and comparing the effects of these winds on an incident can help the FDISO make a difference!

Humidity The FDISO working a large wildland fire will have established communication with a fire behaviorist working the fire as well as a weather forecaster. One important piece of this communication is the relative humidity of the air, because lower humidity means increased fire spread. The FDISO assigned to a structure fire needs to look at humidity somewhat differently. Humidity affects firefighters. In especially dry environments (hot or cold), firefighters become dehydrated quickly by breathing. In high-temperature, high-humidity environ-

WIND PATTERNS FOR TYPICAL THUNDERSTORM

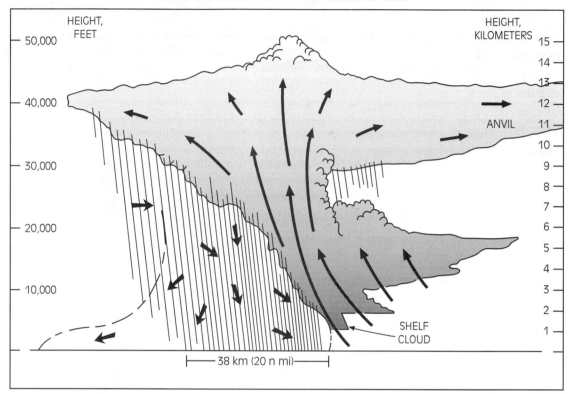

Figure 8-17 *Erratic wind shifts are common during thunderstorms. The FDISO needs to watch weather influences during incidents*

ments, the firefighter becomes dehydrated through profuse sweating in an attempt to cool the human machine (remember metabolic thermal stress?). In cold temperature, high-humidity environments, the firefighter fights penetrating cold (pain) and rapid ice buildup. High humidity can cause smoke to stagnate and hang close to the ground. Prolonged low humidity can cause accelerated fire spread in lumber and other wood products such as shake shingles and plywood. Flying brands can retain their heat longer and fly further in low humidity.

Temperature Which is easier, Launching an aggressive brush fire attack when it is 100°F or maintaining a critical defensive fire stream between two buildings when it is −10°F? Answer: Neither. From the FDISO perspective, temperature needs to be evaluated relative to its effect on firefighter exposure. Acclimation is key here. Firefighters working in International Falls, Minnesota, are quite acclimated to the cold. A firefight in subfreezing temperatures may even seem common. Conversely, firefighters in Yuma, Arizona, are probably accustomed to aggressive operations when it is 105°F. The FDISO needs to consider this acclimation when evaluating the weather and its affect on firefighters. The FDISO should be concerned when firefighters are operating outside of the temperature norm for that region.

Potential for Change or Storms   In most cases, the weather present for a given incident is the weather that everyone has to endure, for better or for worse. At times, however, the duration of the incident or the development of a severe storm during an incident causes significant danger to responders. The FDISO familiar with local weather tendencies can advise the IC of weather indicators that may require a shift in operations. Simple weather spotter guidelines and instructions like the following can help the FDISO warn others.

1. *Watch the sky.* This simple activity is often overlooked. If foul weather is approaching or suspected, find a close vantage point to evaluate cloud patterns and wind activity. A quick phone call to mutual aid agencies upwind may reveal useful information about the approaching storm.

2. *Note 180° changes in wind direction over a short period of time (a few minutes).* A sudden calm should be noted. A wavering smoke column is likewise noteworthy. In most cases, each of these events is a sign of an unstable air mass and may point to a significant weather change.

3. *Be mindful of the potential for a flash flood.* Mentally combine previous rainfall and ground saturation with current rainfall and determine the potential for a flash flood. Evaluate your incident location and whether you are in a low, flood-prone area or along a drainage path.

4. *Developing thunderstorms can produce rapid changes.* These include straight-line winds (microbursts) of 100 mph, hail, and lightning. Tornadic thunderstorms, characterized by a rotation between the rainfree base of a cloud and a forming wall cloud (see Figure 8-17), are especially worthy of the FDISO's attention. A spotted tornado formation in the vicinity or path of an incident should be immediately communicated to the IC and may be grounds for rapid building evacuation and switch to defensive operations. Protective withdrawal from the area may even be considered.

5. *At night, use lightning flashes to help define cloud formations that may be tornadic.* Large hail (1/4" or larger) can also indicate that you are near the area where a tornado will most likely form.

6. *Calculate the distance between you and the lightning.* Do this by counting the seconds between the lighting flash and the thunder. Divide this by five and you have the distance in miles. Thunder that claps less than 5 seconds from the flash means you are in the lightning area! Although lightning can strike anywhere at anytime, firefighters can help protect themselves by becoming the smallest target possible. Stay away from poles, fences, and trees. Avoid standing in water. Suspend operations with aluminum ground ladders (practice solid risk management here).

7. *Snow depth not only makes travel difficult but can hide hazards at the incident.* The weight of snow can stress buildings as well as utility lines, poles, signs, and so forth. Further, ice storms may cause collapse of buildings, power poles and lines, and signs. Hose advancing becomes difficult in deep snow, causing rapid fatigue of firefighters. Driven snow can impair visibility and make smoke reading difficult. Rapid crew rotation and rehab are key.

Having read this chapter, you may think that an environmental reconnaissance is lengthy and complex. Realistically, a skilled FDISO can achieve this survey and evaluation in a matter of minutes. Education, experience, and proficiency speed up this process without sacrificing key observations. Some of the factors and considerations listed in the previous pages may not apply to a specific incident, therefore accelerating the FDISO's recon effort.

Summary

Of all the components of the Incident Safety Officer Action Model, reconnaissance will take most of the FDISO's time. "Recon," as we call it, can be divided into two general areas: environmental and operational. Environmental recon is a study of the many elements we face as emergency responders. The FDISO can frame the environment by first declaring a specific, principal hazard. By defining the principal hazard, the FDISO can focus on specific hazards that are created by the principal hazard. The FDISO uses tools such as the reading of smoke to help further define the principal hazard. Once the principal hazard is defined, the FDISO must determine the stability or integrity of the environment, including observations about building construction, location of fire, and likelihood of change. This integrity survey also includes items such as utilities, vehicles, and access. Environmental reconnaissance also includes a survey of surrounding elements such as topographical layout, entrance and egress, and weather. The FDISO must be able to gauge the effects of weather on the incident and forecast change or any hazards created by the weather.

Review Questions

1. Discuss the difference between environmental and operational reconnaissance.

2. List the three primary areas of environmental reconnaissance.

3. Explain the importance of defining a principal hazard during incidents.

4. What four components of smoke need to be evaluated to predict fire behavior? Explain some examples of each.

5. List three common principal incident hazards and discuss the specific hazards within each.

6. Define four ways to classify the integrity of an incident.

7. List five common electrical equipment items and their associated hazards.

8. Explain the effects of temperature and humidity on an incident. List four weather spotter guidelines that can help forecast weather.

Notes

1. Fire Investigation Report, One-family Dwelling Fire, Three Firefighter Fatalities, Pittsburgh, Pennsylvania, February 14, 1995 (Quincy, MA: NFPA, 1996).

2. Fire Investigation Report, Seattle, Washington, Firefighter Fatalities, January 5, 1995 (Quincy, MA: NFPA, 1996).

Chapter

9

Reconnaissance—Evaluating Operations

Learning Objectives

Upon completion of this chapter, you should be able to:

■ Explain the importance of an action plan to overall incident management.

■ List common missing elements in the support of an incident action plan.

■ List two forms of freelancing and explain the dangers of each.

■ Utilize two methods to determine if fire flow is being met during incidents.

■ Describe the concept of magnet tasks and their inherent danger.

■ Define the four common hazard exposures that lead to injury on the incident scene.

■ Describe the role of the FDISO in crew rehab and list ways to improve rehab efforts.

You've got the building pegged—a three story, ordinary construction tax-payer with a free-burning fire that has taken control of the third floor. Two attack crews have advanced their 1 3/4-inch lines up to the third floor and you are starting to see some signs of steam conversion and knock down. The vent team is just now coming down off the flat roof. A stream of dense, black smoke issues from the new opening. Your first reconnaissance cycle uncovered that no secondary escape existed for the third floor crews. Since then, ladders have been thrown, which give a measure of comfort to the operation.

A rapid intervention team is in place and has been briefed about operations and building considerations. Still, something is wrong, something doesn't quite fit. You watch the smoke coming from the vent hole and the smoke coming from the vent hole again . . . that's it. We aren't meeting fire flow for this fire! Sure, there is some steam and some peripheral knock down, but we aren't stopping the fire. A quick calculation shows that the fire flow should be. . . ."

WORDS OF WARNING

The Incident Safety Officer Action Model arena of reconnaissance comprises most of what the ISO is responsible for in keeping the Incident Commander (IC) informed of safety issues (Figure 9-1). In addition to the environmental reconnaissance items listed in Chapter 8, the FDISO must be keenly aware of operations underway. Conceptually, the FDISO can make the greatest difference at the incident scene in this area. By observing tasks, exposure of crews, tool use, team effec-

THE FIRE DEPARTMENT INCIDENT SAFETY OFFICER ACTION MODEL

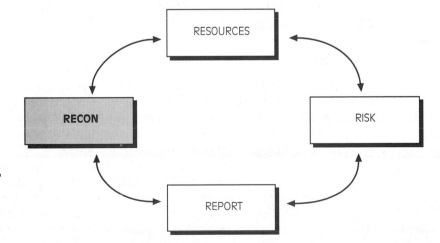

Figure 9-1 *Recon is made up of both environmental and operational surveys.*

■ Note
The operational reconnaissance (recon) task can lead to an FDISO getting sucked into the incident either as a worker or as a freelancer.

tiveness, and other operational events, the FDISO can give focused reports to the Incident Commander as well as evaluate the potential for injury. A word of caution here: The operational reconnaissance (recon) task can lead to an FDISO getting sucked into the incident either as a worker or as a freelancer. These traps do little to further incident safety. In fact, the FDISO who freelances is likely to harm not only him- or herself, but possibly the department's whole FDISO program.

One other caution. Operational recon places the FDISO in a position of evaluating crew effectiveness, which often gives the FDISO an aura of being some kind of quality control specialist. Crew effectiveness needs to be evaluated in the spirit of reducing injury potential and not in the vein of training or skill proficiency. It is the effective FDISO who evaluates the operation and passes on factual observations to the IC regarding what may or may not be effective. Obviously, if the operation is imminently dangerous to life, the FDISO must step in and utilize the authority of the incident commander to stop or alter the operation (discussed in Chapter 11).

What makes up operational recon? Quite simply, it is observations of tasks, teams, and exposure to hazards.

TASKS

A reconnaissance or evaluation of tasks places the FDISO in the unique position to verify that the Incident Commander's action plan is being performed effectively. This can be extremely valuable in not only the management of the incident, but in firefighter safety. In some cases, the FDISO may discover that the IC's action plan may not be appropriate or is creating a risk situation that is endangering firefighters inappropriately. In these cases, the FDISO must present facts and a personally owned concern for safety to help convey sincerity and avoid the appearance of an attack on a decision made by the incident commander. When looking at tasks, the FDISO should consider four points: the action plan itself, the effectiveness of the tasks being completed, any missing elements, and the presence of freelancing.

Action Plan

In essence, the **action plan** is a statement of control objectives and any specific tactical or support actions that support the plan. Generally speaking, the incident action plan should include a basic strategy (rescue, offensive attack, defensive hold, etc.), essential tactical objectives to achieve the strategy, and any support objectives (lighting, rapid intervention crews, rehab, etc.)

More often than not, the action plan is not written down by the incident commander. In these cases, the FDISO should verbally confirm the action plan with the incident commander during the initial confirmation of the FDISO assignment. The FDISO might suggest that the action plan be written if the incident is of a significance where multiple agencies or jurisdictions are involved or when

action plan
the objectives that help achieve the desired strategies for an incident, approved and updated by the Incident Commander as needed

■ Note
The FDISO should verbally confirm the action plan with the incident commander during the initial confirmation of the FDISO assignment.

rapid intervention crew
a dedicated team prepared to effect a rapid rescue of trapped and/or injured firefighters during a defined incident

■ **Note**
NFPA Standards 1500 and 1561 both require the formation of a rapid intervention crew early in the incident if working firefighters are exposed to dangerous environments or tasks.

the incident will see a shift of personnel (including the incident commander), such as long duration or multialarm fires. Many fire departments include a place for the incident action plan on the incident commander's tactical worksheet or status board.

When reviewing an action plan, the FDISO should achieve a sense of the risk profile that the particular incident evokes. The listed objectives should support this. For example, the action plan may list a fire attack objective to aggressively stop the fire on the second floor. In this case, the IC has indicated that we will take calculated risks to save valued property. If the building in this example is a partially demolished or abandoned building, the risk profile may be out of sync.

Another item to be considered when looking at the action plan is whether appropriate plans are in place to support tactical objectives. The best example here is the formation of **rapid intervention crews** (RIC). NFPA Standard 1500 (Fire Department Occupational Safety and Health) and NFPA 1561 (Standard for Fire Department Incident Management Systems) both require the formation of a rapid intervention crew early in the incident if working firefighters are exposed to dangerous environments or tasks.[1] The FDISO should ensure that a rapid intervention crew has been formed and is ready for rapid rescue of other firefighters (Figure 9-2). This readiness should include appropriate personal protective equipment (PPE), self-contained breathing apparatus (SCBA), radio, forcible entry tools, lights, and a life rope. As a habit, the FDISO should also brief the RIC on incident and hazard information including building layout, assignment locations, and fire status.

Task Effectiveness

Once the action plan is known, the FDISO can observe the crews performing those tasks to see if in fact they are being effective. If a crew is not being effective, it may be operating outside the action plan or in a manner that will cause injury. Additionally, the evaluation of crew effectiveness serves as an extension of the IC's eyes. Facts about crew effectiveness can be relayed to the IC during face-to-face updates. There are many ways to measure crew effectiveness. The obvious measurement is that the crew is accomplishing what the Incident Commander believes is being done. Perhaps a more important measurement is the FDISO's perception of whether the task is being performed safely or within an acceptable risk limit.

Picture a crew heading to the roof of a two-story commercial structure with a flat roof. What clues would indicate that the task is being done safely? Most fire officers would expect to see two ways off the roof, a crew of three people who can communicate with the ground or command, a small cache of tools being moved to the roof that includes a powered saw, axe, pike pole or trash hook, rope, and perhaps a charged line. The FDISO may hear the ventilation group or sector supervisor report the results of a cut inspection hole. Finally, the FDISO should see a release of fire gases and smoke within a reasonable time.

Figure 9-2 *The FDISO should ensure that a rapid intervention crew is available. Spending a moment briefing the crew is time well spent. (Photo courtesy of Richard W. Davis.)*

Missing Elements

While performing operational reconnaissance, the FDISO may discover that an essential task is missing. Picture a significant trench rescue operation where crews are attempting a manual dig. Shores are being cut and placed, confined space protocols are being utilized, and crews are being staged for rapid relief. Then it dawns on the FDISO that no thought has been given to the shuttling of dirt—they are stacking it near the hole! This is a safety item that may go unnoticed until the removed dirt gets in the way of the operation.

accident chain
a series of events and conditions that can lead or have led to an accident

Fire officers who have investigated accidents usually discover that a missing element was a trigger to the **accident chain.** This missing element is what the FDISO needs to look for proactively. Skills used to project the incident forward are used. Recollection of training sessions and drills help find missing links when performing recon. Forecasting helps the FDISO find missing tasks. "Forecasting" is a term used in the National Fire Academy curriculum Incident Safety Officer.[2] In this sense, the FDISO is tasked with looking at the future of the incident as it relates to responder safety. A missing element may very well contribute to an emergency.

Some common missing elements include secondary escape ladders, ventilation, lighting, exposure hoselines, backup lines, rapid intervention crews, prestaging of needed equipment, law enforcement backup, fuel for apparatus, medical support, shoring/bracing/cribbing, technical assistance, and zoning (Figure 9-3).

Freelancing

In the business environment, freelancing is viewed mostly as a positive business venture, usually leading to innovation, niche marketing, and entrepreneurial adventures. In the fire service, freelancing is viewed as a dangerous and deadly enterprise. Firefighters have been killed and seriously injured while engaged in

Figure 9-3 *The FDISO needs to watch all operations and try to spot any missing elements.*

a freelance operation, that is, an operation or task that is being performed unknown to the Incident Commander or other working crews. Freelancing is most often attributed to a lone worker although the term can be equally applied to a rogue crew that has determined that another task needs its personal attention. Each scenario is potentially dangerous.

The solo firefighter is probably the deadlier of the two scenarios (Figure 9-4) because of the "what-if" potential. What if the firefighter falls through the floor? What if the firefighter suffers a heart attack? What if the firefighter gets lost? What if the firefighter pushes the fire back on other crews? These "what-ifs" are endless and lead to one conclusion: Nobody will know that the firefighter experienced an emergency or injury. Layer into this scenario that the lone firefighter may get sucked into a situation that requires skills he or she may not possess, or at minimum, requires more than one person to accomplish the task.

The FDISO should keep a close eye on working crews and develop an eye for catching lone workers. Further, the FDISO can apply some basic knowledge of fireground operations and predict situations that lead to freelancing. These include situations where a crew is performing an assigned task and finds that it needs additional equipment. Often, the crew is broken up and one member is sent for the missing tool. This is a "gofer" situation. The incident turns tragic as the gofer becomes lost, is trapped, or is distracted by a more demanding need. Likewise, the person waiting for the gofer to return may look to find ways to continue

Figure 9-4 *The solo worker is the worst and potentially most dangerous form of freelancing.*

to be productive, leading to more what-if hazards. Another freelancing situation can exist when a firefighter is assigned to a seemingly "trivial" task such as monitoring an exposure line or monitoring a positive pressure fan. These firefighters, especially if inexperienced, may search for any excuse to get where the action is.

Similarly, a freelancing crew or team is usually a result of a crew that completed a less-action-oriented task and now wants to get very involved. Often, the task that the freelancing crew is performing is most likely a task that really needs to be done. Unfortunately, the Incident Commander may not be aware of the situation or is aware of the need for the task but has higher priorities in mind for the crew that should have come available after completing the initial assignment. In these cases, the FDISO can catch the freelancing act prior to injury if, and only if, the FDISO has intimate knowledge of the incident action plan and is in communication with the incident commander.

TEAMS

Another component of operational reconnaissance includes the evaluation of assigned teams. Here, the FDISO observes the team and forecasts injury potential in the areas of protective equipment, tool use, team makeup, and rapid withdrawal potential.

Personal Protective Equipment

The fire department incident safety officer who must spend an inordinate amount of time addressing personal protective equipment (PPE) issues will be destined to ineffectiveness (see bunker cop syndrome in Chapter 5). Unfortunately, the failure to address PPE issues could lead to injury. For this reason, the FDISO must still look at PPE issues and ensure that an appropriate level of PPE is maintained by persons exposed to hazards. When violations of PPE requirements are found, the FDISO is best guided to address the issue with the violator's supervisor. Doing so achieves one important goal. The supervisor is given a message that he or she is the person responsible for addressing PPE issues, which will ultimately help switch the focus of the FDISO so that more effectiveness can be realized on scene.

On a more effective level, the FDISO should evaluate scenewide PPE versus individual compliance issues. Here, the FDISO can be proactive in establishing PPE criteria for the duration of the incident, which is especially relevant if the scene requires PPE items that the department does not routinely use. For example, a technical over-the-edge rescue may necessitate the use of fall-arresting devices for rescuers. Another example includes recovery personnel working an especially gruesome aircraft incident where bloodborne pathogen-resistant PPE is more appropriate than standard turnout gear. The FDISO may recommend that both be used depending on the hazards present.

■ **Note**
The FDISO can recommend that certain PPE be doffed because hazards have been mitigated to a degree that the PPE really is not needed.

To the delight of working crews, the FDISO can recommend that certain PPE elements be doffed because hazards have been mitigated to a degree that the PPE really is not needed. Many departments leave the authority to go "packs off" with the FDISO. This determination is usually made with a combination of readings from a carbon monoxide (CO) monitor and the placement of volume fans to increase air flow through the structure. One caution: SCBA may not be required according to the CO monitor, but particulate matter may still be suspended or kicked up as burnt debris is trampled or moved. In these cases, the FDISO may advise that paper filter masks still be required by people working in the "burn."

Tool versus Task

When observing teams, the FDISO can make an evaluation on whether the tools being used by the crew are being effective for the task assigned. This simple evaluation can catch situations that may become injurious, especially if the tool being used is not appropriate for the task.

In some cases, the task being attempted may be beyond the manufacturer's recommended limits for the tool being used. Ladders placed in precarious places can lead to injuries (Figure 9-5). A firefighter using a pike pole as a fulcrum lever-

Figure 9-5 *Using the right tool for the task helps eliminate injury potential.*

■ **Note**
Evaluate the tool
application and deter-
mine the probability
and severity of a
mishap with the tool.

age can easily be injured when the handle fractures. Hydraulic overhaul with master streams can launch debris hundreds of feet with incredible velocity. The trick here is to evaluate the tool application and determine the probability and severity of a mishap with the tool. In some cases, the tool may not be utilized as it is intended, but on-scene firefighter ingenuity is at play and is appropriate for the gained benefit to the operation.

Meeting fire flow with attack fire streams is imperative to a safe and successful fire suppression effort. The FDISO should observe signs that a fire attack is being successful. In cases where no progress is being observed, the FDISO should discern if the attack is failing due to inappropriate fire flow or inappropriate application (access) to the seat of the fire. Failure to make this evaluation may lead to a situation where firefighters are overrun by the fire. This information should be relayed to the Incident Commander. Rapidly expanding steam that overtakes issuing smoke is a good indication that fire flow is being met. Pockets of steam that pale in comparison to the volume of smoke generated may mean that the fire is releasing more British thermal units (BTUs) than the stream can match. In these cases, the FDISO should be quick to communicate with the IC that fire flow is inadequate. Failure to meet fire flow means the attack crew can be overrun by the fire or may be victims of a progressive flashover. Quick calculations of fire flow requirements can help the FDISO determine if the crews have the safety of the appropriate line. On interior fire fights, the use of the Iowa State fire flow formula is appropriate (Figure 9-6). The Iowa State formula estimates the needed flow, in gallons per minute, necessary to knock down a fire in an enclosed area.

Safety
Failure to
meet fire flow means
the attack crew can be
overrun by the fire or
may be victims of a
progressive flashover.

The National Fire Academy adjusted fire flow formula is a good tool to determine if total fire flow available on scene is available for interior attack, exposure protection, and backup (Figure 9-7).

The use of these two formulas will help the FDISO solve the problem of whether the fire flow is adequate for the situation. Remember though, dictating to the IC or other operations officer that more lines are needed might end the FDISO usefulness on the fireground. Relaying observations that crews have not met fire flow underscores the safety concern.

Tool versus team evaluations by the FDISO are often founded in a gut feeling about the potential for an accident. Asking the rhetorical question, "What is the worst thing that can happen with that tool, and what is the likelihood of it happening?" is a good start.

$$NF = \frac{V}{100}$$

where NF = needed flow in gpm

V = volume of the area in cubic feet

100 = is a constant in ft^3/gpm

Figure 9-6 *Iowa state fire flow formula.*

$$NF = \frac{A}{3}$$

where NF = needed flow in gpm

A = area of a structure in square feet (length × width)

3 = constant in ft^2/gpm

Figure 9-7 *National Fire Academy adjusted fire flow formula.*

Team versus Task

Two realities have emerged in modern firefighting: staffing levels have reached critically low levels in many fire departments and firefighters themselves cannot wait for the next exciting or significant incident. These two realities present an ambiguous environment for the incident safety officer. On one hand, the incident will have fewer people to attack the incident than we had in the past. On the other hand, those tasks that are exciting or different will be a magnet to firefighters anxious to help. When evaluating the team or teams working a task, the FDISO must determine if the task is being tackled with too few or too many firefighters. If too few people are working the task, the FDISO may cause the IC to assign more people to the task in order to eliminate injury potential. The converse may also need to be addressed. If too many people are addressing the task, injury may result due to congestion and immobility of the team (Figure 9-8).

Figure 9-8 *It is just as dangerous to have too many people perform a task as it is to have too few.*

■ Note

A magnet task is the one that everyone wants to be involved with, like the forcing of a vehicle door with a hydraulic rescue tool or the removal of a live victim from a debris pile.

!Safety

Magnet situations create a negative risk in that a single mishap is rapidly compounded by the proximity of many firefighters.

"Magnet" tasks are usually subject to an overload of firefighters. A magnet task is the one that everyone wants to be involved with, like the forcing of a vehicle door with a hydraulic rescue tool or the removal of a live victim from a debris pile. In communities with few fires but many firefighters, it is not uncommon to see a 1 3/4-inch attack line being advanced into a residential kitchen fire with no less than five firefighters attached to it! While at a Midwest fire department, a visiting fire officer witnessed a dozen firefighters ventilating a pitched roof on a single story, 1,300-square-foot home. The one firefighter with the chain saw did a wonderful job in spite of the eleven other assistants giving advice!

These magnet situations create a negative risk in that a single mishap is rapidly compounded by the proximity of multiple firefighters.

In other cases, the right team for the task is not an issue of too many or too few, but rather an issue of the right expertise for a given task. The advent of multiple specialty teams, and the classic division of engine, truck, and rescue companies have created a virtual cornucopia of team combinations that may or may not make an incident safe. The FDISO should examine the task at hand and confirm that the right set of skills are being applied by the appropriate team.

Rapid Withdrawal and Rapid Intervention

Essential to the FDISO size-up of team versus task and tool versus task is the visualization of rapid withdrawal of the assigned team from the task. For example, it is one thing to have two ways off a roof, it is yet another to visualize the roof crew bailing off the roof in the face of a collapse. Are the ladders within easy reach? Are there enough ladders for the number of people on the roof? Are the ladders in a collapse zone?

In large I-Zone fires, the FDISO should evaluate the plan to get everyone out of the way of a potential blowup (Figure 9-9). The effort to get five engine crews and three tender crews off a hillside subdivision in the face of blowup is critical. Likewise, the FDISO should ask if safe zones are established in a highrise fire so that attack crews can retreat and compartmentalize themselves from the fire.

Once again, the planning and readiness of rapid intervention crews (RIC) are important. The New York City Fire Department (FDNY) assigns a truck company as a Firefighter Assistance and Search Team (FAST) on working incidents. The FDNY FAST concept has been quite successful and has been adopted by many fire departments nationwide. The Loveland Fire and Rescue Department (Colorado) uses a cascading RAT (Rapid AcTion) group to address immediate intervention. The RAT group is the fourth company to be formed on scene (following first-due, incident command, and truck functions). If a second attack group is needed, the RAT groups moves in and a company out of staging becomes the RAT group. This has worked well for motivating crews and helps keep crews from being disappointed that they have to sit and wait for a firefighting emergency.

The FDISO should brief the officer of the FAST, RIC, or RAT group of any hazards associated with teams or tasks as well as any observations or concerns regarding the reconnaissance of the environment. Most RICs appreciate the FDISO

■ Note

The FDISO should brief the officer of the FAST, RIC, or RAT group of any hazards associated with teams or tasks as well as any observations or concerns regarding the reconnaissance of the environment.

Figure 9-9 *The FDISO should inquire about predesignated escape plans and safe areas during wildland and I-Zone fires.*

approaching them and advising them of the situation, building construction features, and any noted hazards. Should a mishap occur, the RIC will be more prepared. Large-scale incidents may require more than one RIC. Multiple alarm fires, incidents spanning a large geographical area, and incidents with a high-risk potential to many firefighters are all candidates for multiple RICs.

EXPOSURE TO HAZARDS

Chapter 4 presented numerous hazards that firefighting crews can be exposed to. The FDISO applies that information in evaluating the exposure of firefighters to hazards. From the FDISO perspective, these exposures fall into four key areas: exposure to fire/heat/smoke, potential for trips and falls, proximity to collapse zones, and effectiveness of rehabilitation efforts.

Exposure to Fire/Heat/Smoke

Modern personal protective equipment has been designed to give the firefighter a survival chance should exposure to fire products become severe. Unfortunately, this same protection has created a situation where firefighters are comfortable in getting close to a hostile fire environment (Figure 9-10). The time that an incident safety officer has to read the fire, determine the crew exposure, and warn crews of the need for withdrawal has been compressed to a mere few seconds because of modern PPE. The FDISO should be ever mindful of this time compression and therefore be keenly alert of changing fire behavior signs. Mastering the art of reading smoke is paramount (see in Chapter 8 under Defining the Principal Hazard).

The FDISO should look at the fire conditions and determine the potential for a crew to get trapped. The dynamics of fire are such that a crew working above a fire can easily be cut off from an exit. Likewise, a simple assignment to protect an exposure could lead to a fire entrapment. For example, Figure 9-11 shows a crew protecting an exposure as the fire is drafting (rolling) inward. If the exposure crew moves toward the dead end, a split second change in the fire draft—a result of a new ventilation opening elsewhere in the structure—will cause the crew to be burned.

❗Safety
● The time that an incident safety officer has to read the fire, determine the crew exposure, and warn crews of the need for withdrawal has been compressed to a mere few seconds because of modern PPE.

Figure 9-10 *The time that a FDISO has to read a fire and warn of potential flashover is being compressed.*

Figure 9-11 *A split-second change in fire behavior is possible here. The crew should not move toward the fire.*

Trip/Fall Hazards

■ **Note**

The number one type of firefighter injury is strains and sprains that result from slips, falls, and jumps.

■ **Note**

When fall hazards cannot be removed, the FDISO may need to simply remind the working crews that they are in a "fall haven" and to please take it slowly or cautiously.

The number one type of firefighter injury is strains and sprains that result from slips, falls, and jumps.[3] Therefore, it stands to reason that the FDISO should pay particular attention to the potential for these injuries at the incident scene. At a recent house fire on a cold, wintry day, a fire attack pumper was supplying two attack lines. From underneath the pump, a slowly increasing ice slick was forming. Seeing that the assigned crews were still in an aggressive posture, the FDISO envisioned someone slipping on the ice while concentrating on tool retrieval or moving about the apparatus. The FDISO grabbed absorbent material (like kitty litter) from a compartment and spread it over the ice slick, then placed an orange traffic cone smack-dab in the middle of the ice. An injury was undoubtedly prevented.

Figure 9-12 shows a typical overhaul operation following the extinguishment of an attic fire that consumed the roof of a house under construction. How many trip or fall hazards can you count? How many of the trip or fall hazards could be eliminated? Figure 9-13 shows just a sampling of the potential for trip and fall injuries for this scenario. In cases where fall hazards cannot be removed, the FDISO may need to simply remind the working crews that they are in a "fall haven" and to please take it slowly or cautiously. The FDISO can make a huge difference in reducing incident scene slip and fall injuries.

Figure 9-12   *Can you count the trip or fall hazards in this typical overhaul operation?*

Collapse Zones

Incident commanders want information on when and where a building is going to collapse. Actually, that may be the first question the IC asks of the FDISO. The FDISO needs to analyze the building using the seven-step process presented in the Building Construction section of Chapter 4 (Figure 9-14), then report collapse potential findings to the IC.

Further, the FDISO needs to communicate collapse zones to all working crews. Once this is done, the FDISO can watch crews and police these collapse zones (Figure 9-15). Can a crew operate in a collapse zone? To answer this question, the FDISO needs to evaluate the task being performed and combine the evaluation with solid risk management. Occasionally, the situation may require a crew to work in a collapse zone. Crews removing debris from a person trapped under a previous collapse may be an example of this situation. Likewise, crews

> **! Safety**
> The FDISO needs to communicate collapse zones to all working crews. Then the FDISO can watch crews and police these collapse zones.

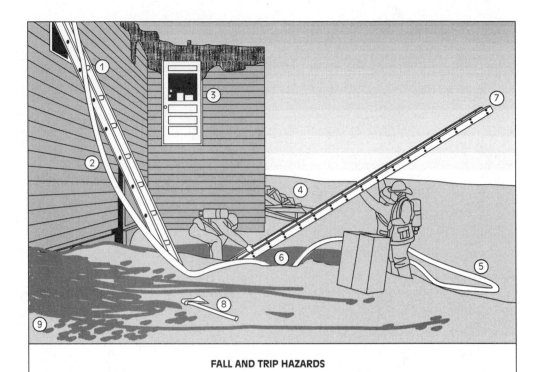

FALL AND TRIP HAZARDS

1. UNSUPPORTED LADDER
2. UNATTENDED HOSE AND NOZZLE
3. DOORWAY WITH NO LADDER OR STAIR
4. CONSTRUCTION DEBRIS
5. HOSELINE UNDERFOOT

6. WORKING IN DITCH OR ACCESS CUT
7. WILL WALL SUPPORT LADDER? (ROOF IS GONE)
8. FIRE AX
9. CONSTRUCTION GRADE (NOT SMOOTH, MANY RUTS)

Figure 9-13 *Did you spot more than nine?*

¡Safety
● If crews have worked to the point of physical and mental exhaustion, an injury is looming.

■ Note
The goal of incident rehabilitation is to prevent injuries with timely and appropriate crew rest, hydration, and nourishment.

attempting to pull a ceiling in order to access an attic or "cockloft" fire are working in the collapse zone. Judgments on structural integrity become crucial. These judgments should be based on the understanding of the effects of fire and gravity on building construction and not just physical signs of impending collapse.

Crew Rehabilitation

It does not take a physician to tell if a firefighter has had enough (Figure 9-16). Usually if crews have worked to the point of physical and mental exhaustion, an injury is looming. The goal of incident rehabilitation is to prevent injuries through timely and appropriate crew rest, hydration, and nourishment.

While NFPA 1500 clearly states that the incident commander will consider the circumstances of the incident and initiate rest and rehab procedures,[4] the

Figure 9-14 *The seven-step building analysis process presented in Chapter 4 will help the FDISO proclaim collapse zones to the crews.*

Seven-Step Process to Analyze and Predict Building Collapse

1. Classify the type of construction.
2. Determine the degree of fire involvement.
3. Visualize load imposition and load resistance.
4. Evaluate time as a factor.
5. Determine the weak link.
6. Predict the collapse sequence.
7. Proclaim collapse zones.

Figure 9-15
If collapse zones have been communicated, there will be no surprise when collapse does occur.

■ **Note**
NFPA 1500 clearly states that the incident commander will consider the circumstances of the incident and initiate rest and rehab procedures.

FDISO needs to evaluate these efforts and see if they are, in fact, being successful in meeting the rehab goal.

Chapter 4 covered human performance and keyed on the importance of proper methods to reduce thermal stress and properly hydrate and fuel the firefighter. When on scene, the FDISO should evaluate the rehab process and check to see essential rehab elements are in place. These elements include timely rotation, adequate rest and recovery, sufficient rehydration, and balanced nutrition.

Figure 9-16
Firefighters overdue for rehab are high risk for injury.

■ **Note**

NFPA 1500 states that the highest level of available emergency medical care shall be on standby during "special operations" and that this emergency medical care at a minimum is BLS.

Additionally, the FDISO should ensure that a minimum of Basic Life Support (BLS) emergency medical service is available. NFPA 1500 states that the highest level of available emergency medical care shall be on standby during "special operations" and that this emergency medical care at a minimum is BLS.[5] Ideally, the rehab effort will include medical checks by qualified BLS or Advanced Cardiac Life Support (ACLS) personnel (Figure 9-17).

These med checks can be quick: a brief look at skin color, rapid check of pulse, and general impression about level of fatigue. If any signs or symptoms of excessive fatigue, heart stress, or thermal stress are present, the EMS provider can mandate additional medical assessment or treatment as well as mandating a longer rest period for the firefighter. The Loveland Fire And Rescue Department's rehabilitation policy gives the EMS provider the power of the incident comman-

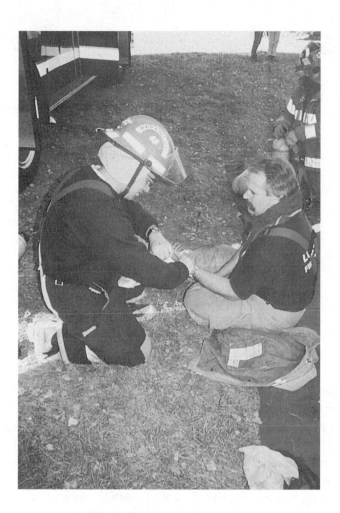

Figure 9-17 *Rehab efforts should also include provisions for quick medical checks of working firefighters.*

der to mandate additional rest, hydration, or further medical treatment, including transportation to a definitive medical facility.

BACK TO THE INCIDENT SAFETY OFFICER ACTION MODEL

Remember, reconnaissance, both environmental and operational, is but one of four components of the Incident Safety Officer Action Model. Reconnaissance is ongoing and must remain cyclic. The other components, resources, risk, and report, must also be addressed.

Summary

Operational reconnaissance focuses not so much on the surroundings of an incident but on the many tasks and function that firefighters employ. The FDISO looks at tasks, teams, and the exposure of personnel. Additionally, the FDISO should evaluate the rehab effort. When observing tasks, the FDISO is most focused on making sure that the action plan is being carried out in a way that is expected of the Incident Commander. Often, missing action plan elements are obvious, leading to a potential injury situation. Freelancing in forms of rogue crews or the classic individual effort should be found and eliminated. Teams themselves should be observed to make sure the right team for the task and the right tool for the task are being utilized. The exposure of crews to incident hazards is also of concern. The ability of a FDISO to read a hazard, predict its next phase, and discover the relationship of a crew is time compressed. Rapid reconnaissance is a must. Trips and falls are the number one cause of firefighter injuries so the FDISO must give this hazard particular attention. Finally, crew rehabilitation must be evaluated for effectiveness. At any one incident, the FDISO may make over a dozen recon trips to continue the cyclic-thinking Action Model.

Review Questions

1. Explain the importance of an action plan to overall incident management.

2. List four common missing elements in the support of an incident action plan.

3. List two forms of freelancing and explain the dangers of each.

4. Using the National Fire Academy fire flow formula, figure the fire flow for a two-story building, 55 feet × 120 feet with a 25% involved fire on the first floor.

5. Describe the concept of magnet tasks and their inherent danger.

6. Define the four common crew exposures that lead to injury on the incident scene.

7. Describe the role of the FDISO in crew rehab and list ways to improve rehab efforts.

Notes

1. NFPA 1500, Fire Department Occupational Safety and Health Program, Chapter 6, Section 5 (Quincy, MA: NFPA, 1997 edition), p. 20.

2. Incident Safety Officer, Student Manual, FEMA/USFA/NFA-SO-SM, (Emmitsburg, MD, 1994) pp. 5–11. October 1994.

3. Karter and LeBlanc, "1996 U.S. Firefighter Injuries," NFPA Journal 90, no. 6 (1996): p. 68.

4. NFPA 1500, Fire Department Occupational Safety and Health, Section 6-6, Rehabilitation During Emergency Operations (Quincy, MA: NFPA, 1997), p. 20.

5. NFPA 1500, Fire Department Occupational Safety and Health, Section 6-4.5 (Quincy, MA: NFPA, 1997), p. 19.

Chapter

10

Risk Evaluation and Management

Learning Objectives

Upon completion of this chapter, you should be able to:

- List the three influences that determine acceptable risk.
- Explain the acceptable risk guidelines outlined in NFPA standards.
- Apply the frequency/severity matrix to help prioritize risks at an incident.

As part of your department's standard procedure, an incident safety officer responds with your crews to other jurisdictions for mutual aid requests. You are now following a pumper and aerial from your department that are responding to just one of these requests. Enroute you recall the information given so far—a three-block square, four-story warehouse is about 20% involved and the incident has escalated to its third alarm. Upon arrival, the requesting agency's chief radios your group to set up on the west side, secure a water supply, and prepare for an aerial master stream operation. The incident commander (IC) rejects your offering to help with safety. Your personal size-up reveals that many crews are still inside, and two ladder pipes are being played into upper floor windows on the involved south side. The alarm warnings go off in your head: "Offensive and defensive operations being done simultaneously?" Your crew urges you to do a face-to-face with the IC and find out what's going on. This can't be right! As you walk that way, you try to rationalize a risk situation that would allow these competing strategies. You almost trip with a studder-step as you realize the answer: A firefighter rescue might be underway!

RISK MANAGEMENT OVERVIEW

The Incident Safety Officer Action Model includes a component simply titled "Risk" (Figure 10-1). Compared to the rest of the Action Model, the risk portion is more an overhead issue and not as detail-oriented as the many topical issues addressed in resources or recon. In the risk portion of the Action Model, the fire department incident safety officer (FDISO) is reminded to look at the big picture of any risk taking that crews may assume. In some cases, the FDISO looks at the strategy and tactics employed by the IC. Other times the FDISO looks at the inherent risks of a specific crew and its actions. This review requires the FDISO to look at many levels of risk management. These levels are best expressed positionally, that is, each position or role on the incident has certain responsibilities in terms of risk management. The United States Fire Administration released a text called *Risk Management Practices in the Fire Service* that best defines the risk management responsibilities for various incident roles (Figure 10-2).

■ **Note**

In the risk portion of the Action Model, the FDISO is reminded to look at the big picture of any risk taking that crews may assume.

In all positional cases, the FDISO is actually making an evaluation of whether the risk being taken is acceptable or unacceptable. Although this evaluation seems black and white, the FDISO must consider the grayness of many factors to make the ultimate decision if a risk is acceptable. These factors include community expectations, fire service standards, department values, and situational application of risk management priorities.

Determining the appropriateness of risk taking is perhaps the most difficult decision that the FDISO has to make at the incident scene. In most cases, the Incident Commander will have already established risk boundaries for working crews. For example, the IC who has pulled interior crews and deployed exterior

THE FIRE DEPARTMENT INCIDENT SAFETY OFFICER ACTION MODEL

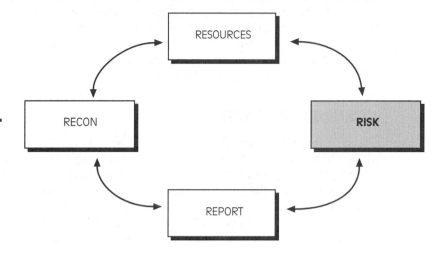

Figure 10-1 *The Action Model reminds the FDISO that "risk", as in risk management, is an essential component in injury reduction.*

Figure 10-2
Everyone has a role in risk management. Source: U.S. Fire Administration Risk Management in the Fire Service, *FA-166, Emmitsburg, MD, Dec. 1996, p. 76.*

United States Fire Administration Suggested Roles of Risk Responsibility

Roles and responsibilities   Every position in the emergency response system has a role in operational risk management, as the table that follows indicates.

Role	Responsibilities
Incident Commander	Is expected to make strategic decisions based on risk acceptance or avoidance. Decisions are incorporated in the selection of offensive or defensive operations, a major component of the strategic plan for the incident.
Sector and company officers	Are expected to supervise tactical operations based on risk acceptance/avoidance in the situations they encounter within the areas they supervise. Their determinations must be consistent with the direction provided by the incident commander. They must keep the incident commander informed of any situations they encounter that may have an impact on the strategic plan.
Individual firefighters/ EMS personnel	Might be called on to make personal decisions about risk acceptance/avoidance when no officer is present, which must also be consistent with the strategic plan and with departmental policies. They also need to keep their officers informed of any significant information.
Incident Safety Officer On-Scene Risk Manager	Is the risk management consultant for emergency operations. The incident safety officer is an advisor to the incident commander and should provide an overview of the situation specifically directed toward identifying and evaluating safety concerns.

defensive fire streams has, in essence, determined that firefighters may no longer take the risk of fire attack within the structure. In other cases, the line is not so clear. In these cases the FDISO has to make the difficult decision of whether a specific task, strategy, or action is worth the risk of injury. Some have expressed this as deciding risk versus benefit. In order to help make risk judgments, the FDISO should look to the following factors.

COMMUNITY EXPECTATIONS

Most firefighters accept the notion that they may have to risk their own life to save a life. In most cases, the community served by the firefighter expects that of the firefighter and the fire department. This basic premise or understanding is what separates the firefighting profession from many others and gives the profession respect within a community.

Firefighters who swear to uphold the values of the community—to protect life and property—must draw on courage and, in some cases, bravery to achieve this. This same courage and bravery must be laced with a heavy dose of prudent judgment so that the situation does not unnecessarily harm the firefighter.

In cases where a firefighter dies protecting property, the taxpayer may question if prudent judgment was present. Clearly, the community does not expect firefighters to die needlessly. Therefore the IC and the FDISO must establish a pace, approach, or set of guidelines so that an unnecessary injury does not occur.

FIRE SERVICE STANDARDS

■ Note
NFPA standards specifically address risk management concepts related to the handling of emergency operations.

NFPA standards specifically address risk management concepts related to the handling of emergency operations. These concepts are spelled out quite clearly and should be incorporated into the incident command function and the incident action plan. These concepts are actually acceptable risk guidelines from which risk decisions can be made. They are:

1. Activities that present a significant risk to the safety of members shall be limited to situations in which there is a potential to save endangered lives.

2. Activities routinely employed to protect property shall be recognized as inherent risks to the safety of members, and actions shall be taken to reduce or avoid those risks.

3. No risk to the safety of members shall be acceptable when there is no possibility to save lives or property.[1]

Most fire officers would subscribe to these risk principles. These concepts can be shortened and made easy to remember with the following:

Risk a life to save a life.

Take a calculated and weighted risk to save valued property.

Take no risk to save what is lost.

The interpretation of these principles is still wide open. Some further discretion on the part of the FDISO may be required. For example, the first statement says "Risk a life to save a life." Notice that it does not say "sacrifice" a life to save a life. Throwing firefighters at a rescue may not be the answer. Likewise, the statement is not plural. At some point, the IC or FDISO must determine how many firefighters should be at risk to save a given life. A recent trench rescue in the eastern United States claimed four responders' lives (one police officer and three firefighters). Each thought they could enter the below-grade hole and save the original victim and subsequent rescuer, creating a tragic chain of sacrifice.

When discussing a "weighted and calculated" risk to save valued property, the FDISO should first determine what "valued property" is. Although definitions may vary, it seems prudent to attach a qualifier here. Valued property can best be defined as property, that if lost, would cause harm to the community. Examples may include a hospital, a significant employer, a utility infrastructure, or a place with historical significance. In these cases, the FDISO should calculate the risk (possibility and severity of an injury), then weigh safety factors in favor of the firefighter. For example, an interior firefight at a County Clerk's office may be extended because of the potential for loss of irreplaceable records. The FDISO may recommend that additional engine companies be ordered for backup or that emergency roof and ceiling bracing be installed to prolong the time the firefighters can fight or start salvage operations.

DEPARTMENT VALUES AND SKILLS

When determining an acceptable or unacceptable risk, the FDISO must consider what is commonplace and accepted by the department. In one department, the aggressive attack of an interior fire may not be acceptable until ventilation efforts are underway. The department next door may decide that it is acceptable to start a fire attack before ventilation is accomplished. In this example, one department has placed an extreme value on ventilation. In one small, rural Kansas fire department, *no* interior fire attack is allowed on anything other than an incipient fire because the department simply does not have the equipment and training for that level of risk.

The FDISO must evaluate a given operation and consider if the situation fits the organization's "normal" way of handling the incident. A conflict can easily arise here. The department may have embraced an inherently unsafe way of doing a task. A good example is the department that allows forms of freelancing due to insufficient on-scene staffing or the department that routinely launches an interior fire attack before four people have assembled on scene. In these cases, the FDISO must be extra alert and should try to build a bridge toward a safer operation. These cases will try the FDISO and present a challenge to creatively reduce injury potential.

The FDISO should recognize situations where crews are attempting to perform a skill for which the members have never prepared. The rise-to-the-task attitude of most firefighters creates this situation. In these cases, the FDISO should encourage technical assistance, a slower pace, and increase of zoning or backup.

SITUATIONAL APPLICATION

Risk management was presented in Chapter 2. From this presentation, a frequency and severity chart was used to help prioritize risk abatement efforts (Figure 10-3). This graphic can be easily applied to specific incidents where unusual circumstances or multiple hazards are clouding situational risk.

Elsewhere in this book, the questions have been raised that can help the FDISO determine if a risk is acceptable. These questions include:

What is the worst possible thing that can happen here?

What is the likelihood of it happening?

How severe will the injury be?

Can any intervention be employed to reduce risk?

Have our people ever tried this before and what was the result?

The FDISO can draw from many avenues to determine if a given situation presents an acceptable risk scenario. Knowledge, sound judgment, experience, and wisdom are paramount in making risk decisions. Many FDISOs prepare for

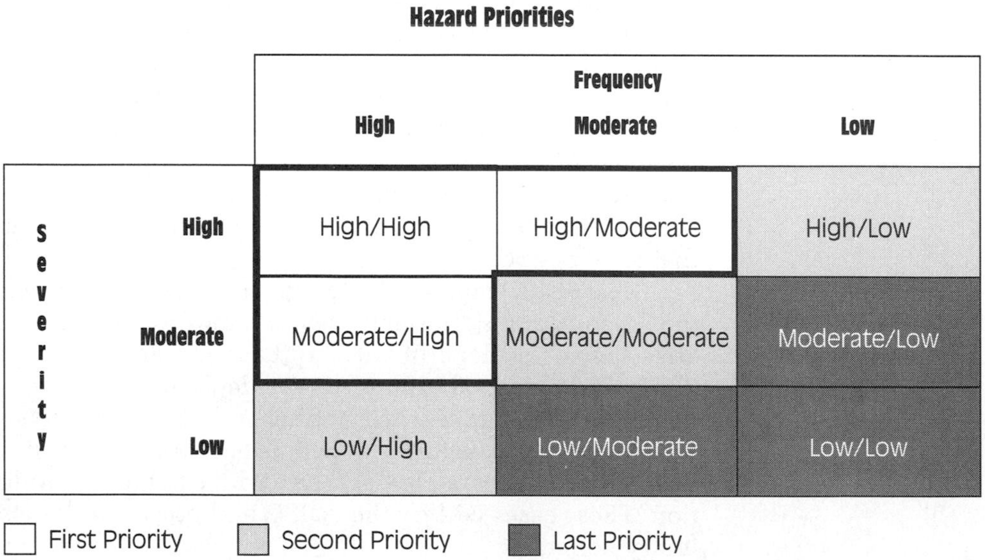

■ **Note**
Knowledge, sound judgment, experience, and wisdom are paramount in making risk decisions.

Hazard Priorities

		Frequency		
		High	**Moderate**	**Low**
S e v e r i t y	**High**	High/High	High/Moderate	High/Low
	Moderate	Moderate/High	Moderate/Moderate	Moderate/Low
	Low	Low/High	Low/Moderate	Low/Low

☐ First Priority   ▨ Second Priority   ▨ Last Priority

Figure 10-3 *Using the severity and frequency chart can help the FDISO prioritize hazards.*

this situational decision making by reading the many accident investigation reports generated for firefighter duty deaths. This vicarious experience has, in many cases, prevented another firefighter death.

If the FDISO is uncomfortable with the risks being taken but cannot quite justify or articulate his or her concern, then that, in itself, is reason to visit with the incident commander. Sometimes the FDISO may find it effective to ask the incident commander how she or he feels about the situation—sort of a reality check. This is an example of being a consultant and helps in the team approach to managing risks.

Summary

Deciding the appropriateness of risk taking is perhaps the most difficult judgment that the FDISO must make. The Incident Safety Officer Action Model forces the FDISO to address this issue. In most cases, external influences predetermine the definition of an acceptable risk. An example of this is the community that expects its firefighters to risk their own life to save that of another. Likewise, NFPA standards present a general, acceptable risk-taking outline. The FDISO must draw on these influences as well as his or her own judgment, values, and knowledge to determine if a risk being taken is acceptable. The use of frequency/severity matrix may help this process.

Review Questions

1. List the three influences that outline acceptable risk.

2. Explain the acceptable-risk guidelines listed in NFPA standards.

3. Using the frequency/severity matrix, compare and contrast the priority of risk taking for common fireground tasks with and without a rescue operation.

Note

1. NFPA 1500, Fire Department Occupational Safety and Health Program, 6-2.1.1 (Quincy, MA: NFPA, 1997), p. 18.

Chapter

11

Reporting Conditions and Taking Action

Learning Objectives

Upon completion of this chapter, you should be able to:

- List the four main elements of the report component of the Incident Safety Officer Action Model.
- Detail the items that should be communicated to the incident commander during safety status updates.
- Explain the importance of FDISO action documentation.
- List and explain the essential elements of a safety briefing.
- Explain the authority given the FDISO to take action.
- List the two classifications of hazards and detail strategies to take action on these hazards.

By far, it's the biggest incident you've ever been involved with. There are six mutual aid departments here, over eighty firefighters, and you have three people working for you as a safety sector. You've decided to have two of your assistant incident safety officers work on reconnaissance and the other monitor communications and work as your aide. Frankly, you knew that one day you'd have the "big one," but never did you think it would be a downed airliner in this housing area. The multiple fires, resource juggling, and search and rescue effort have stressed everyone but it appears that the incident management system is in place now and crews are working with a keen sense of purpose.

The Incident Commander wants to have an update and planning meeting with you, the Operations Chief, and the Planning Chief. As you assemble your notes and head over to the meeting, you mentally package your report. You want to make sure to mention the hazards that your two reconnaissance people have pointed out, especially upgraded PPE needs and decon. You also want to relay a concern about crew rotation and try to get a feel for the incident duration. Critical incident stress teams have been ordered and you'd like to get a feel from Operations on how best to incorporate defusing into crew rotation.

The Incident Commander listens to everyone's reports then gives a nod for an updated action plan. You add a few last comments regarding the new action plan. The Incident Commander looks over to you and says, "I need you to develop two different safety briefings, one for the relief firefighters coming in to help with recovery and one for the whole host of agency reps that will get involved. Next, I want you to prepare a structural status report. Get help from the building inspectors. And we need to tag buildings as safe for entry or not. Finally, I'd like you to meet with the Logistics Chief I'm going to assign and make sure that she's square on rotation, diffusion, and all the zones."

As you take your notes away from the meeting you reflect on the enormity of this operation. Thoughts flow through your head about how you're going to get everything done that you've been asked to do, not to mention all the things that safety should continuously do. Then it strikes you. Of all the fallout and potential outcomes of this sad, catastrophic event, the one thing you don't want now is a hurt firefighter. You roll up your sleeves and dig in.

■ **Note**

The basic premise of the report element is that of sharing hazard concerns with the IC.

REPORTING CONDITIONS

So far, Section 2 has discussed three of the four elements of the Incident Safety Officer Action Model. The last element to be covered is that of report (Figure 11-1). The basic premise of the report element is that of sharing hazard concerns with the incident commander (IC). Being the last element covered in this book does not

THE FIRE DEPARTMENT INCIDENT SAFETY OFFICER ACTION MODEL

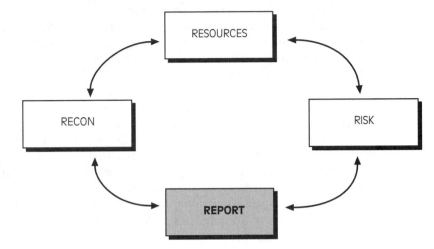

Figure 11-1 *Report is the catchall component of the FDISO Action Model.*

necessarily mean that report is the last component addressed on an incident. Reporting may very well be the first item addressed by the FDISO. An example of this is the FDISO who arrives, receives an action plan update from the incident commander, then delays reconnaissance to discuss preplan or weather information with the Incident Commander in order to avert potential injuries right away.

Typically, however, the report part of the Action Model comes later in the incident and is repeated often. The elements of a good report include the careful and prioritized relaying of found hazards as well as a summary of action plan effectiveness. The report phase is also a good place to consider contingency planning, assisting with the development of an updated action plan, and the creation of notes, sketches, safety briefings, and other records.

Hazard Priorities and Concerns

It is hard to imagine anyone finding no hazards at an incident: The nature of the fire fighting and emergency rescue profession is just that—hazardous. Training proficiency, modern protective equipment, standard operating procedures, high-tech apparatus and equipment, and solid tactics and strategy combine to reduce these hazards to what we hope to be an acceptable level. The FDISO will, however, find risks that pose a threat of injury to firefighters even in the company of the aforementioned. Each risk has an associated degree or potential for injury. This area is where the FDISO is required to exercise judgment and present a report that captures the priority of each hazard facing incident personnel. The frequency/severity chart presented in Chapters 2 and 10 will assist in assigning priorities.

At times, the FDISO will get a bad feeling regarding an incident. Interestingly, some of these ill feelings or perceptions have been communicated to investigators assigned to collect witness statements following a significant firefighter injury or fatality. If these feelings present themselves prior to an accident, they should not be neglected. Although the most important FDISO reporting task is to articulate facts, observations, and judgments, it is also important, to a lesser degree, to communicate that you, the Incident Safety Officer, do not feel comfortable with the incident circumstances. If for no other reason, this intuition will sharpen senses and make you and the IC more attentive.

In terms of raw information, the FDISO should, at minimum, communicate a summary of the other action model elements: resources, reconnaissance, and risk. Specifically, the report should include an evaluation of fire behavior (or primary hazard), a summary and confirmation of building construction and integrity, and an honest appraisal of operational effectiveness. The Incident Commander can take this information and move forward with an established action plan or make needed changes.

Action and Contingency Planning

At large-scale incidents, the Incident Commander may assign the FDISO to check with the plans section to review a proposed action plan. In order to be effective in this capacity, the FDISO should ensure that the required elements of the Action Plan are in place. These elements include incident objectives, resource allocation, desired outcomes for a given operational period, and any safety or hazard information. Often, it is the FDISO who brings in what-ifs that can lead to contingency planning. Attention to details such as rapid intervention teams, weather factors, traffic and access, escape plans, and rehab efforts add to the credibility of the FDISO and the establishment or reinforcement of safety values for planning personnel.

At relatively small-scale incidents that still include an FDISO, the FDISO should develop several contingencies should an unforeseen event or circumstance present itself. At a recent incident where a roof had partially collapsed on workers during construction, the FDISO asked the IC what the evacuation plan was should the remainder of the roof start to collapse during rescue. The IC asked the FDISO to come up with a plan and communicate it to the working crews. Many departments preplan a procedure for just this example. These procedures generally include an immediate evacuation signal (airhorns, radio alert tone, etc.) followed by a personnel accountability report **(PAR),** which is a crew-by-crew radio report to determine if all firefighters are accounted for. Specific to the incident, predesignated safe areas and evacuation routes should be established.

Documentation

The FDISO documents an incident in two ways. First, the FDISO should assist in the collection and evaluation of documents that assist with the handling of the

■ Note
Attention to details such as rapid intervention teams, weather factors, traffic and access, escape plans, and rehab efforts add to the credibility of the FDISO.

PAR
personnel accountability report; crew-by-crew radio report to determine if all firefighters are accounted for

■ **Note**

The FDISO should make it a point to jot down a few notes regarding his or her own actions and observations during the incident.

incident, that is, preplans, material safety and data sheets (MSDS), and blueprints. Second, the FDISO should contribute to the documentation of incident efforts and required reporting.

Ideally, all safety officer actions should be documented as the incident progresses, but it is often an impossibility due to the compressed nature of action time during especially stressful or rapidly changing situations. Practically speaking however, the FDISO should make it a point to jot down a few notes regarding his or her own actions and observations during the incident. These notes will assist in the overall reporting of the incident as well as the preparation of a **postincident analysis** (PIA) or critique. In the case of a firefighter injury or fatality and/or equipment accident, documentation will have to be filed by the FDISO. PIAs, accident reporting, and documentation are discussed more in Chapter 12.

postincident analysis
formal or informal
critique session
following an incident

Some departments use tactical planning boards and checklists to aid in the management of an incident. These provide a quick and easy format for documenting the FDISO's actions. A sample incident safety officer checklist that highlights the FDISO Action Model is included as Appendix C.

Safety Briefings

Pointing out unsafe conditions and communicating hazards require constant vigilance and attention by the FDISO. One FDISO task, the creation and delivery of safety briefings for assigned crews, can help crews focus on unsafe conditions and hazards prior to their deployment. Typically, time does not allow for the assembly of a crew and delivery of a safety briefing prior to task engagement. Mostly, we rely on preincident safety briefings—that is, safety briefings during drills and training sessions that will cover essential behaviors and cautions. Some incidents, however, are quite conducive to the delivery of a safety briefing. Sustained incidents, major campaigns, mutual aid operations, highly complex or technical situations, hazmat incidents, and operations during terrorist or unusual circumstance situations are all candidates for the inclusion of a safety briefing prior to firefighter action.

■ **Note**

Sustained incidents, major campaigns, mutual aid operations, highly complex or technical situations, hazmat incidents, and operations during terrorist or unusual circumstances are all candidates for the inclusion of a safety briefing prior to firefighter action.

A well-planned safety briefing is ideally delivered by the FDISO although a written briefing can be delivered by any supervisor or sector, branch, group, or division officer. A good safety briefing includes a general summary of the action plan and the accepted risk-taking level of the operation. More specifically, a safety briefing should include:

- Specific tactical assignments or objectives
- Summary of the anticipated conditions and required PPE
- Chain of communication and status update reminder
- Known or suspected hazards
- Access and evacuation routes
- Firefighter emergency procedures

- Collapse zones or other hot, warm, and cold zones
- Rehab and medical attention plan

It may seem that this list could be time-consuming and awkward, but in most cases, these items can be covered in just a few short sentences. Complex or highly technical situations may require more time to accomplish the briefing, but the safety mind-set and specificity of the safety briefing is worth the time and effort expended.

TAKING ACTION

Undoubtedly the incident safety officer will encounter a situation where action, ranging from a simple advisory to an immediate order to evacuate an area, is necessary to prevent injury. The authority for an FDISO to take action should be predetermined at the department level. Nationally speaking, two important safety documents agree that any person filling the safety officer position should have the situational authority to stop, alter, or suspend actions. This authority is contained in NFPA 1521 (and repeated in NFPA 1561, Fire Department Incident Management System[1]) as well as in the Occupational Safety and Health Administration (OSHA) Hazardous Waste Operations and Emergency Response (HAZWOPER) Regulation 29 CFR 1910.120.

NFPA 1521 reads:

2-5.1   At an emergency incident where activities are judged by the incident safety officer to be unsafe or to involve an imminent hazard, the incident safety officer shall have the authority to alter, suspend, or terminate those activities. The incident safety officer shall immediately inform the incident commander of any actions taken to correct imminent hazards at the incident scene.[2]

OSHA HAZWOPER reads:

(q) (3) (viii)   When activities are judged by the safety official to be IDLH (Immediately Dangerous to Life and Health) and/or to involve an imminent danger condition, the safety official shall have the authority to alter, suspend, or terminate those activities. The safety official shall immediately inform the individual in charge of the ICS (Incident Command System) of any actions needed to be taken to correct these hazards at the emergency scene.

■ Note
Regardless of the action taken by the FDISO, the FDISO must report to the incident commander.

In recognition of these two quotations, the FDISO Action Model includes an "Act" component (Figure 11-2). As can be seen, any of the Action Model elements can lead to an observation where action is necessary. Regardless of the action taken by the FDISO, the FDISO must report to the incident commander. This recognizes the role of the FDISO as a staff officer subordinate to the incident commander.

FDISO ACTION LEADS TO REPORT

Figure 11-2 *Any of the Action Model components can cause the FDISO to take action. Any action, however, leads the FDISO to report to the incident commander.*

imminent danger
situation that is immediately dangerous to life and health

potential danger
situation that, given time, proximity, or subtle change, can cause injury

The action taken by an FDISO should fit the recognized hazard. Hazards should be classified as either an **imminent danger** or a **potential danger.** With these classifications comes a set of action options for the FDISO. Remember, though, regardless of the action, the FDISO must report the action to the IC.

Imminent Danger

Imminent danger is best defined as any situation that is immediately dangerous to life and health and will likely cause injury or death to a firefighter or crew. In most cases, this situation will be obvious and self-announcing based on information and conditions known by the FDISO (Figure 11-3). Crews exposed to the threat may not have the same information or point of observation known by the FDISO. In these cases, the FDISO has three action choices. First, he or she may immediately stop the crew from performing its assignment. Second, the FDISO can choose to have the crew retreat its position. In these cases, the crew may still perform its assigned task—just from a safer location. Finally, the FDISO may order the crew to alter its task. An example of an alteration would be a change in an attack position for fire streams or the shutdown of a fire stream until other crews have retreated. The alteration of an apparatus position (such as an aerial parked in a collapse zone) is another example of altering a task to eliminate an imminent threat.

In each of these options, the FDISO may be put in a position to explain the reason for the action. The FDISO should be prepared to articulate and justify the action taken. The justification should be based on fact, realistic anticipation, and solid risk management foundations. Further, the manner in which the action is

Figure 11-3 *Once an imminent threat is perceived by the FDISO, he or she should immediately communicate the hazard along with specific instructions. (Photo Courtesy of Richard W. Davis.)*

taken should be positive and confident. As stated before, any action taken by the FDISO needs to be reported to the IC.

Potential Danger

In the case of potentially dangerous situations, the FDISO should take one of three actions. Most commonly, the FDISO simply advises the exposed crew of the concern (Figure 11-4). This advisory is best delivered face-to-face with the crew although the FDISO may not have that opportunity due to the distance to the crew or the atmosphere the crew is working. If face-to-face communication is not an option, radio transmission is the second choice. Crew advising is particularly effective because it is simply that—advisory in nature—as opposed to directive.

Most company officers or supervisors welcome the added information, or in some cases, expertise from the FDISO. The crew officer is left to lead and manage his or crew based on the shared concern.

Another option is that of offering suggestions to the team facing the potential danger. In essence, this is an extension of the consultant role that the FDISO has with the incident commander. In cases where the FDISO has no method to communicate with a crew facing a potential danger, the FDISO may choose the last option, which is to consult directly with the incident commander (or other

Figure 11-4 *Most potential hazards can be handled with a simple crew briefing.*

operations chief or leader), thereby sharing concerns with the very person who has control responsibility with working crews.

In each of these options, it is important to report the option chosen, as well as the message delivered. Once again, the FDISO should be prepared to defend any concern or action in a positive and confident manner.

BACK TO THE FDISO ACTION MODEL

Reporting conditions, providing for documentation, creating safety briefings, and taking action are all part of the report component of the Action Model. Continuing the Action Model cycle—resources, reconnaissance, and risk—is imperative.

Summary

The report component of the FDISO Action Model is actually the catchall component. During the report phase, the FDISO should brief the IC on all pertinent hazards and any suggestions or actions that may be warranted. These safety status updates should include all essential safety information as well as a status of crews, building construction (integrity), and results of judgments in the areas of resources, recon, and risk. The FDISO should also use the report phase to forecast action and contingency planning, provide incident documentation and to develop safety briefings. A safety briefing is an excellent way to inject a safe attitude prior to assignment of crews. The safety briefing should include many key components regarding PPE, access, emergency procedures, and the like. Hazards can be divided into two main classes: imminent danger and potential danger. OSHA regulations and NFPA standards give the FDISO authority to alter, suspend, or terminate activities that pose an imminent danger to firefighters. In potential danger cases, the FDISO most often meets with the exposed crew and relays concern. Regardless of the action the FDISO takes on hazards, the IC must be immediately informed.

Review Questions

1. List and explain the four main elements of the report component of the Incident Safety Officer Action Model.

2. What specific items should be communicated to the incident commander during safety status updates?

3. Explain the importance of FDISO action documentation.

4. List and explain the essential elements of a safety briefing.

5. Explain the authority given the FDISO to take action.

6. Compare and contrast the two classifications of hazards and detail strategies to take action on these hazards.

Notes

1. NFPA 1561, Fire Department Incident Management System, Command Staff, 3-2.2.2 (Quincy, MA: NFPA, 1995).

2. NFPA 1521, Fire Department Safety Officer, 2-5 Authority of the Incident Safety Officer (ISO) (Quincy, MA: NFPA, 1997).

Chapter 12

Postincident Responsibilities of the FDISO

Learning Objectives

Upon completion of this chapter, you should be able to:

■ Explain the factors that lead to injuries during postincident operations.

■ Discuss the role of the FDISO in informal and formal postincident analysis.

■ List the six specific items that the FDISO should comment on during PIAs.

■ Explain the role of the FDISO in accident investigation according to NFPA standards.

■ List the five parts of the accident chain.

■ List and explain the three steps of accident investigation.

Three crews are inside the one-story home pulling ceilings and finishing up the overhaul process following an attic fire. As the incident safety officer, you begin addressing postincident relaxation and fatigue issues that typically lead to accidents. As you leave the rehab area, satisfied the rehydration effort is going well, you notice a sudden gathering around the incident commander. Upon approach you hear one crew leader say to another, "Man am I glad to see you. I went out the front and I thought you were still under." The other officer replies, "No, we bailed out the side. It came down quick."

"That was a close call. Man were we lucky!"

For every one serious firefighter injury, there are more than six hundred near misses or close calls that could have easily been serious.[1] At times, firefighters wear these close calls as a badge of courage and use exaggerated tales to fuel firehouse coffee talk. Other times, firefighters involved with a near miss trivialize or minimize the brush with injury or death. Often, the closer the firefighter came to serious injury, the more he or she minimizes the story telling, perhaps indicating that the event really got the firefighter's attention. At what point, however, does the Incident Safety Officer need to follow up on a near miss and work toward the prevention of a similar event that may not have such "lucky" consequences?

Ideally, the lessons learned from *any* near miss should be included in training and used for ongoing efforts to avoid similar situations in the future. Often, the war stories that are told from close calls are invaluable tools in the teaching at new-firefighter academies nationwide. The key to making these lessons productive is an accurate portrayal of the situation's facts and actions. The only way for this to happen is to collect information quickly and accurately. The FDISO's responsibility to prevent injuries through careful incident reconnaissance places him or her in the best position to collect and document incident activities for postincident analysis or critiques. Further, the FDISO can use this information to begin an investigative process if the unfortunate injury or fatality has occurred. This chapter explores the responsibilities and duties of the FDISO for postincident activities, postincident analysis, and accident investigations.

■ **Note**
The lessons learned from any near miss should be included in training and used for ongoing efforts to avoid similar future situations.

■ **Note**
Postincident injuries seem almost ironic in a profession where aggressive and calculated risk taking is a hallmark.

POSTINCIDENT ACTIVITIES

Although no accurate documentation or data exists, many injuries occur while crews are packing up to leave an incident. Common postincident injuries include strains, sprains, and being struck by objects. The causes of these postincident injuries seem almost ironic in a profession where aggressive and calculated risk taking is a hallmark. Let us examine a few causes of postincident injuries and steps the FDISO can take to reduce these seemingly minor injuries.

Postincident Thought Patterns

One cause of postincident injuries has to do with the little-studied notion of postincident thought patterns, essentially inattentiveness. After especially difficult, unusually spectacular, or particularly challenging incidents, firefighters tend to reflect on their actions. The replay of the incident starts almost instantly when the order is given to "pick up." This introspection is normal (Figure 12-1). The switch from activities requiring brain power to an activity that is so routine as to be dull is a hard one to make. Herein lies the problem.

The FDISO should circulate among the pickup effort and keep an eye out for inattentiveness. Signs may include faraway stares, robotlike actions, and firefighters who stop and look about as if they have forgotten their task. Simple reminders or jocularity can help regain focus and reduce injury potential. One method to reduce the impact of these thought patterns is take a time out and have everyone gather for a quick incident summary and safety reminder. These huddles can be effective for everyone or just small groups (Figure 12-2). Even a casual-coach approach that emphasizes the need to stay alert and not fall into an injury trap can be useful.

■ Note
Simple reminders or jocularity can help regain focus and reduce injury potential.

Chemical Imbalance

Even the most successful rehabilitation program cannot prevent firefighters from experiencing fatigue and mental drain. With the end of an incident, especially one requiring major physical effort, comes the relaxation of the firefighter's mind,

Figure 12-1
Postincident introspection is normal but may lead to inattentiveness and injury.

Figure 12-2 *Calling a huddle before incident cleanup is the best opportunity to remind firefighters of lingering injury threats.*

which, in turn, starts shutting down protective chemicals that stimulate performance. The adrenaline rush is over and the firefighter's metabolism returns to a "repair" state. This causes a mental slowdown that can lead to unclear thinking and resulting injuries.

Another way to see this chemical imbalance and mind relaxation is to look at a firefighter's tools from a layperson's point of view. The layperson would literally have to concentrate on carrying an axe, pike pole, chain saw, roof ladder, or hose length in order to not hurt him- or herself or anyone around. Consider also doing these tasks in bulky, restrictive clothing and heavy boots. The firefighter does these things after incredible energy bursts under frightening conditions. Familiarity plays a certain role, but concentration is still required. If the mind has been taxed, the body has been fatigued, and the signal to "relax" has been given, yet the concentration required to do the task has remained the same, the potential for injury rises.

Whether the issue is chemical imbalance or postincident thought patterns, the FDISO needs to stay alert and try to pick up signs of potential injury and take steps to "awaken" crews.

POSTINCIDENT ANALYSIS (PIA)

Tailboard talk, after-action report, critique, slam session, PIA, incident review, and Monday morning quarterbacking are all labels the fire service has attached

■ **Note**

The successful fire officer always learns something from every working incident he or she is involved in.

■ **Note**

NFPA 1500, 6-8.2: The fire department incident safety officer shall be involved in the postincident analysis as defined in NFPA 1521.

■ **Note**

NFPA 1521, 4-8.1: The incident safety officer shall prepare a written report for post incident analysis that includes pertinent information about the incident relating to safety and health issues. The incident safety officer shall participate in the post incident analysis.

to that formal and informal reflective discussion that fire officers use to summarize the successes and failures of handling an incident. The successful fire officer always learns something from every working incident he or she is involved in, frankly because no two incidents present the same set of circumstances. Although every firefighter involved with an incident has a viewpoint or opinion regarding those circumstances or on how an operation went, only the Incident Commander and the Fire Department Incident Safety Officer have the perspective of an incident overview. From this perspective the FDISO should contribute officially to the postincident analysis.

NFPA 1500 acknowledges this: "6-8.2 The fire department incident safety officer shall be involved in the postincident analysis as defined in NFPA 1521, Standard for Fire Department Safety Officer."[2]

NFPA 1521 goes on to say, "4-8.1 The incident safety officer shall prepare a written report for post incident analysis that includes pertinent information about the incident relating to safety and health issues. The incident safety officer shall participate in the post incident analysis."[3]

In order to maximize the effect the FDISO has on a postincident analysis, it is important to discuss the essential philosophy of postincident analysis as well as FDISO issues surrounding PIAs. Finally, we look at a simple process to ensure that appropriate information is covered by the FDISO for the PIA.

Philosophy

The FDISO should approach any formal or informal postincident analysis with an attitude of positive reinforcement for safe habits and an honest, open desire to prevent future injuries. In most cases, the postincident analysis is nothing more than a discussion of what went right and what we would do differently next time. This may sound simple, but it is often hard to achieve, especially in light of a close call or the occurrence of a significant operational mistake that easily could have led to an injury.

When an operational mistake has been made, the FDISO should first consider the likelihood and severity of an injurious outcome. If the concern warrants, the FDISO should employ a philosophy of discovery. The FDISO can approach discovery from a fact-finding point of view. By asking a few questions about the operational environment or about the general feeling or pulse of the incident, the FDISO may help crews contribute a sense of concern or an acknowledgment of the error. Occasionally, this approach will not work. Crews may spend great amounts of energy explaining their actions in an attempt to justify. In these cases, the FDISO may take the approach of relaying a personal account of the thought processes that led to a belief that crews might be injured. If this is accomplished with a communicated understanding of the crew's point of view and a sense of caring, a message will be sent. At all cost, confrontation should be avoided.

While the general approach of postincident analysis is to look back to an incident, it must not be forgotten that the overriding goal is to look forward to the future. Rather than calling the information at a postanalysis "feedback," perhaps we should call the information "feed forward."

FDISO Issues

Because of the FDISO's task of monitoring an incident for potentially unsafe situations, the FDISO brings many valuable observations to the postincident analysis. Just looking at the items addressed in the Incident Safety Officer Action Model—resources, environmental and operational reconnaissance, risk evaluation, and reporting, it is easy to see the value of the shear quantity of input the FDISO could have. It is important to note, though, a postincident analysis is a time for crews to share and reflect and take home a message. A long dissertation from a FDISO can easily negate any message. The FDISO should comment on some key issues, however, including the following:

General Risk Profile of the Incident The FDISO can share the overall picture from a risk management point of view. Items such as risk-gain or benefit, pace, and impressions about appropriateness of risks taken can be discussed. If a situation developed that placed a crew at risk, the FDISO may find it valuable to call on the crew to relay their thoughts or perceptions. These observations may have to be built upon so that value is gleaned by all.

Effectiveness of Crew Tracking and Accountability The FDISO can yield to an accountability system monitor for some of this information. Observations about crew freelancing (working in conflict of the action plan), individual freelancing (working without a partner), and reinforcement about successful tracking should be shared.

Rehabilitation Effectiveness Very seldom does the FDISO have to be "processed" through rehab. So how then can the FDISO comment on the effectiveness of rehabilitation? Simply, the FDISO can share observations regarding pace, energy and focus trends throughout the incident, and the duration or rotation of work efforts. If injuries resulted during the incident, an investigation is likely. But for postincident purposes, some exploration of rehab as a contributing factor may be discussed.

Personal Protective Equipment (PPE) Use As stated often in this book, it is best if the FDISO shuttles individual PPE concerns to the company officer or crew leader. Overhead issues of PPE can, however, be addressed by the FDISO. As an example, the choice to do overhaul without self-contained breathing apparatus (SCBA) may have been premature. Likewise, a four-gas monitor may have been used to make the decision to go packs-off and use simple dust masks. These decisions can be discussed and reinforced where appropriate.

Close Calls Obviously, the circumstances surrounding a near injury should be detailed from all participants' point of view. The FDISO should contribute but should reserve judgment until all sides have been shared. An actual investigation may be warranted.

Injury Status If no injuries have been reported, this fact should be reinforced; most likely, it is good practice and procedure that leads to having no injuries. If this belief is not shared by all, then the role of luck should be discussed. This discussion could very well expand safety awareness for the next incident. If an injury did occur, firefighters will want an update. The FDISO should be cautious here. Medical confidentiality, investigation results, and other issues may limit the amount of information that can be shared. In these cases, the FDISO should keep the discussion centered on the efforts underway to take care of the injured as well as on an overview of the investigative process.

If a firefighter injury was significant or a firefighter fatality was involved, the FDISO should use the PIA as a tool to listen to firefighters. Often, firefighters may appear to be focusing blame when, in reality, they are venting, displacing stress, or even grieving. Any of these reactions may be a signal that critical incident stress teams be notified.

■ **Note**

Often firefighters may appear to be focusing blame, when in reality they are venting, displacing stress or even grieving. Any of these reactions may be a signal that critical incident stress teams be notified.

■ **Note**

More significant incidents should utilize a prepared, formal PIA within a day or two from the actual incident.

Process

A postincident analysis can take on many forms. Individual department procedure should guide the FDISO on the best process to follow to ensure that the FDISO's observations are noted. Some general guidelines, however, can assist the FDISO in preparing to contribute to a PIA.

First, the FDISO should ascertain from the Incident Commander if he or she wishes to host a formal or informal PIA. Most routine incidents can be analyzed informally just prior to releasing crews from the scene or just after cleanup at the fire station. This procedure is especially effective if no significant operational issue has been raised. More significant incidents should utilize a prepared, formal PIA within a day or two from the actual incident. If a firefighter fatality is involved, a formal PIA may be delayed until the investigation is complete. Whether formal or informal, the FDISO should employ a few simple steps to make each an opportunity to increase everyone's ability to make solid risk decisions and prevent future injuries. Let us take a look at these simple steps.

On Scene Before crews pick up to leave the scene, make it a point to check in and say a few words (Figure 12-3). Also take this opportunity to ask a few questions. One technique that is successful is to ask crews if they noticed any hazards that you, the FDISO, may have missed. Another simple, caring question is to ask if everyone on the crew is feeling OK or has received any minor injuries that you should know about. A parting positive comment regarding the crew's effort should always be included.

Figure 12-3 *The FDISO should make a point of checking in with crews to get a sense of their perspective of the incident.*

Documentation At minimum, the FDISO should document a quick summary of the hazard issues that were discussed with the Incident Commander or any crew (Figure 12-4). This documentation should be included with a summary of building construction features, any unique features of the operational environment, as well as an incident time line. A quick call to the dispatch center can help you record times. Chronological succession of events is often questioned during PIAs. These questions can be clarified if the FDISO has documented a time line.

If a formal PIA is scheduled, the FDISO should spend more time with documentation. It is also recommended that the FDISO confer with the Incident Commander prior to the PIA session to help avoid displaying differing points of view of the IC and FDISO, which, in front of crews, can have the effect of dissing safety. Remember the forward focus as you prepare for a formal PIA.

Trend Spotting As an FDISO, you may find a recurring problem or concern. In these cases, take the time to jot down some thoughts then share them with a supervisor, training officer, or the department's health and safety officer. One point of consideration here. Many people can spot and articulate problems; it is the exception to present a problem along with some reasonable solutions.

ACCIDENT INVESTIGATION

Another Fire Department Incident Safety Officer duty is that of incipient accident investigation. Often the person to begin an accident investigation following a fire-

■ **Note**

Often the person to begin an accident investigation following a firefighter injury, fatality, or equipment mishap is the FDISO.

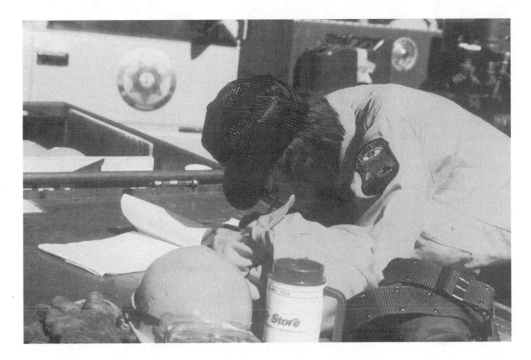

Figure 12-4 *Quick incident documentation is essential for the FDISO. The FDISO perspective and documentation will improve the quality of a postincident analysis.*

■ **Note**

NFPA 1521, 4-7.2: The incident safety officer shall initiate the accident investigation procedures as required by the fire department.

■ **Note**

NFPA 1521, 4-7.3: In the event of a serious injury, fatality, or other potentially harmful occurrence, the incident safety officer shall request assistance from the health and safety officer.

fighter injury, fatality, or equipment mishap is the FDISO, by nature of his or assignment as a command staff member. NFPA 1521, Standard for Fire Department Safety Officer (1997 edition), outlines the duties of the FDISO in regard to accident investigation.

4-7.2   The incident safety officer shall initiate the accident investigation procedures as required by the fire department.

4-7.3   In the event of a serious injury, fatality, or other potentially harmful occurrence, the incident safety officer shall request assistance from the health and safety officer.[4]

An accident investigation is one of the first steps in avoiding future injuries and deaths. Often the results of the investigation can lead to a change of unsafe situations, habits, or equipment not only for the originating department, but for departments nationwide. This vicarious learning from other's mistakes is essential. The Hackensack, New Jersey, tragedy is a perfect example.

In Hackensack, the local fire department was attempting to extinguish an advanced fire in the truss space (cockloft) of an automobile repair shop. The roof collapsed, trapping and eventually killing five firefighters. One accident report cited numerous factors including failure to recognize the danger of collapse of truss-involved fires, command dysfunction, and failure to have a firefighter accountability system. These findings became the impetus for change for fire departments around the country.

Figure 12-5 *The accident triangle shows that for every one serious injury, there are thirty minor injuries and more than six hundred close calls. Close calls should also be investigated.*

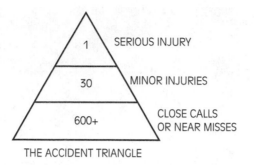

THE ACCIDENT TRIANGLE

Close calls or near misses should also be investigated. The reason for this is discussed in the introduction of this chapter. Consider also that the widely recognized accident triangle (Figure 12-5) suggests that for every one serious accident, there are thirty times as many property damage accidents, and more than six hundred near miss or practice accidents. Frankly, this should scare all of us and provide motivation to take lessons from these practice accidents.

It is easy to see the benefit and importance of a thorough accident investigation. The well-prepared FDISO can truly make a difference in injury reduction through competent investigative skill.

Introduction to Accident Investigation

Accidents are the result of a series of conditions and events that lead to an unsafe situation resulting in injury and/or property damage. Many call this series of conditions and events the "accident chain." The investigation of an accident is actually the discovery and evaluation of the accident chain. The chain includes five components (Figure 12-6):

1. Environment, including physical surroundings such as weather, surface conditions, access, lighting, and barriers.

2. Human factors, including components of human (or social) behavior. Training, use of or failure to use recognized practices and procedures, fatigue, fitness, and attitudes are included as human factors.

3. Equipment, including PPE, limitations and restrictions of equipment, maintenance and serviceability, and appropriateness of application. Some may argue that the misuse of a piece of equipment is actually a human factor.

4. Event is the intersection of the foregoing components. Something had to bring those components together in such a way to create an unsafe or unfavorable condition.

THE ACCIDENT CHAIN

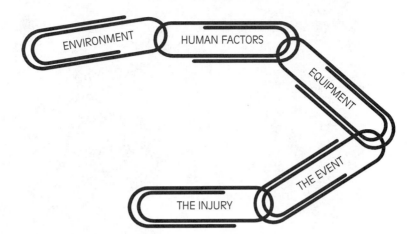

Figure 12-6 *Accident investigation is actually the discovery and linking of the accident chain.*

5. Injury, the last part of the chain, deals with the injury (or property damage) associated with the accident. A near miss or close call is an accident without physical injury. For the sake of the accident chain, this injury can be presumed.

Ideally, the FDISO should stop a potential accident by eliminating one of the elements in the chain. That is the focus of the FDISO Action Model presented in the preceding chapters.

Investigation Issues

The FDISO should be aware of some issues and concerns that will arise regarding the involvement of the FDISO in accident investigation. One of the biggest concerns is that of liability.

Picture this: A firefighter is seriously injured while working a commercial structure fire where an incident safety officer has been appointed (Figure 12-7). In the ensuing investigation, the question arises that if a safety officer was present, should an injury have occurred? The incident safety officer obviously did not do his or her job. This same question leads to doubts about the objectivity of the incident safety officer in doing the investigation of that incident. Did the FDISO cover up anything in an attempt to refocus blame away from the FDISO? This scenario may seem far-fetched, but it is a reality in our ever-litigious society. How does the FDISO perform his or her safety task on scene as well as perform an honest, meaningful investigation following an accident? The answer is simple: Do both with due diligence.

Due diligence is a legal phrasing for the effort to act in a reasonable or prudent way, given the circumstances, with due regard to laws, standards, and

■ Note

Due diligence is a legal phrasing for the effort to act in a reasonable or prudent way, given the circumstances, with due regard to laws, standards, and accepted professional conduct.

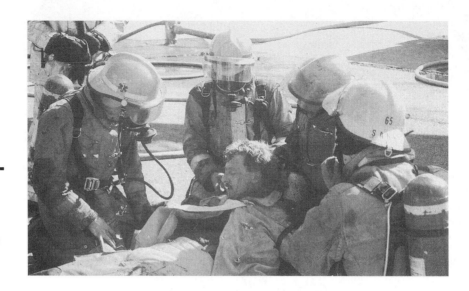

Figure 12-7 *The occurrence of a serious injury on the incident scene presents the FDISO with many issues and concerns.*

accepted professional conduct. If the FDISO acts in a prudent manner, that is, he or she uses the Incident Safety Officer Action Model, takes steps to eliminate or communicate hazards, and works within established standards (National Fire Protection Association) and laws (Occupational Safety and Health Administration regs), then the FDISO has taken significant steps in reducing his or her liability as FDISO. Added to this is a long-standing legal principle of discretionary function. Simply speaking, discretionary function recognizes that certain activities require a value judgment among competing goals and priorities. In these cases, nonliability exists (*Nearing v. Weaver*).[5]

Another issue the FDISO must be aware of in an accident investigation is involvement of outside agencies with an interest in the accident. It is not uncommon to have state and/or federal OSHA officials, labor group investigators, insurance investigators, and law enforcement officials involved in a significant injury or death investigation. In many cases, these agencies can help the FDISO; most likely, however, an investigation that has reached this magnitude signals the end of the FDISO's need to lead or even participate in the investigation.

The Investigative Process

So exactly where does the FDISO begin to investigate an injury or mishap? There are many investigative models to chose from, but the most common approach is a simple three-step approach.[6]

Step 1: Information Collection Numerous items of information should be collected following an incident. These can be divided into a few categories that need to be gathered.

- Incident Data. Included are factual information such as incident number, chronological time of events, weather conditions, apparatus assigned, personnel assigned, and documented benchmarks (primary search complete, incident under control).

- Witness Statements. This information may be difficult to gather and assistance may be required (law enforcement officials can usually help). Here, an attempt is made to gain as many perspectives as possible (Figure 12-8). It is important to keep the witness speaking in facts, but it is also important to gain a sense of perspective from which the witness was operating. Remember, so much of what a firefighter does requires rapid judgment and application of a decision.

- Scene Sketches or Diagrams. Accuracy is an issue here—try to be as precise as possible. Quick hand sketches work well for apparatus/hose/crew placement as long as measured distances are included so that a more precise drawing can be rendered later.

- Photographs or Video. If you noticed video footage being taken during an incident, attempt to gain these from the videographer. Media sources may be helpful. Follow-up video or still photography can help document the accident.

Figure 12-8 *The FDISO should support an accident investigation with many witness reports.*

- Physical Evidence. Protective equipment, damaged equipment, or other physical forms of evidence should be retained (Figure 12-9). Once again, law enforcement officials are a good source of expertise in the collection and documentation of physical evidence.
- Existing Records. Equipment maintenance records, policy and procedure manuals, training records, and such will be useful when analyzing the factors leading to the accident (Figure 12-10). At times, this analysis will involve an extensive search and may require going back many years to make the discovery that led to the injury. Likewise, the research may reveal that the proper maintenance, training, and so forth was in place.

Step 2: Analysis and Reconstruction In step 2, the FDISO reads through the accumulated data and separates facts, perceptions, unknowns, and determines the need for more information. At times, irrelevant data will need to be discarded. Once the information has been analyzed, the FDISO can reconstruct the accident. The reconstruction should be based on known facts, yet acknowledging the judgments and perceptions of involved crews. Once this is done, the accident chain will become apparent. Using the accident chain, the FDISO can move to the next step.

Step 3: Recommendations The chain of conditions and events that led to the accident indicate areas of concern that should be identified. Most often, concerns fall into the areas of equipment, policy and procedure, or personnel (training, attitude, fit-

Figure 12-9
Protective equipment and other physical evidence needs to be retained, marked, tagged, and identified.

Figure 12-10 *Years of documentation may have to be reviewed to reconstruct an accident cause.*

ness). From here, the FDISO can develop some possible solutions to prevent the accident from happening again. Important to note here—there may be a tendency to focus on one solution. Be inventive and force yourself to develop more than one solution.

After multiple solutions have been developed, evaluate each alternative and focus on the approach that you believe would best address the concerns in order to prevent a future reoccurrence.

Notice that nowhere in the accident investigative process are the words "blame" or "discipline" used. This is important. Placing blame or recommending discipline has a tendency to close minds and erect barriers to effective FDISO use. If the FDISO is to remain effective, it is best to state recommendations in the form of future accident prevention. The FDISO should address cases where a complete disregard for safe practice was noted (a form of negligence) with a supervisor or chief officer and allow the department, as an administrative function, to handle the issue. Although there may be some backlash and tension regarding the FDISO, most safety-conscious firefighters and officers will applaud the actions of the department and the FDISO.

Accident investigation is not a fun task, yet it is vitally important to the reduction of future injuries. If the FDISO can focus on this good intent, the investigation will make a difference.

Summary

The FDISO plays a key role in preventing injuries during an incident. As crews begin to pick up, the injury potential still exists. This is compounded by postincident thought patterns (introspection) and the possibility of chemical imbalance that can affect concentration. The FDISO needs to remind firefighters of this phenomena in order to prevent postincident injuries. The FDISO is also in a key position to provide information to postincident analysis reports. Specifically, the FDISO should report on the specific safety items he or she has witnessed during the incident. Following an incident injury, the FDISO may be faced with some serious issues regarding liability and blame. By exercising due diligence, the FDISO can avoid these issues. Regardless, the FDISO must begin an accident investigation. By understanding the accident chain and applying a three-step investigative process, the FDISO can discover and reconstruct the accident. With this information, the FDISO can help prevent future injuries.

Review Questions:

1. Explain the factors that lead to injuries during postincident operations.

2. Compare and contrast the role of the FDISO in informal and formal postincident analysis.

3. List the six specific items that the FDISO should comment on during PIAs.

4. Explain the role of the FDISO in accident investigation according to NFPA standards.

5. List and explain the five parts of the accident chain.

6. What are the three steps of accident investigation? Detail key elements of each step.

Notes

1. D.G. Jones, Accident Investigation Analysis, *Health and Safety for Fire and Emergency Service Personnel* 8, no. 9 (September 1997).

2. NFPA 1500, Fire Department Occupational Safety and Health Program (Quincy, MA: NFPA, 1997).

3. NFPA 1521, Fire Department Safety Officer (Quincy, MA: NFPA, 1997).

4. NFPA 1521, Fire Department Safety Officer (Quincy, MA: NFPA, 1997).

5. *Nearing v. Weaver,* 295 Ore.702,670 P.2d 137, 143 (Ore 1983) in T. Callahan and C. W. Bahme, *Fire Service and the Law* (Quincy, MA: NFPA, 1987).

6. R. H. Hopkins, *Accident Investigation, Health and Safety for Fire and Emergency Service Personnel* 5, no. 9 (September 1994).

Appendix

A

Sample Incident Safety Officer Standard Operating Guideline (Selection, Training, and Duties)

PURPOSE

To develop a system that provides a dedicated, on-call Fire Officer to respond to designated and significant incidents in order to serve the Incident Commander as a Safety Officer.

RESPONSIBILITY

It shall be the Fire and Rescue Department's responsibility to select and train eligible officers to serve as Fire Department Incident Safety Officers (FDISOs). It shall be the selected Fire Officer's responsibility to adhere to the training and procedural requirements of this document.

PROCEDURE

General Program Guidelines

1. The department shall select and train five to eight fire officers (career and volunteer mix) to serve as available Fire Department Incident Safety Officers (FDISOs). Ideally, one career officer from each shift and three to four volunteers would constitute an effective mix.

2. To be eligible for selection, a career or volunteer member must have at least 5 years of fire service experience, with at least 2 years experience as a line officer (Lieutenant or above) and be currently serving as an officer.

3. Serving as an FDISO is voluntary. Members wishing to be selected as an FDISO agree to complete the training program, attend required proficiency training, and serve at least 2 years as an FDISO.

4. Selected Officers who complete the FDISO training shall design a system to ensure that one of the designated Fire Department Incident Safety Officers is available at all times. As a guideline, an off-duty career member trained as an FDISO should be available using the same guidelines as the shift recall system.

5. Career members that are FDISO trained shall not serve as an FDISO while on duty for a scheduled shift position, unless no other FDISO is available and the Incident Commander makes this assignment.

6. FDISOs completing the initial training program will be provided a "Safety Officer" vest, portable radio (career members), and clipboard/metal file. If the member no longer serves as an FDISO, changes from officer status, or fails to complete training requirements, the member shall return FDISO equipment to the department.

Training Requirements

1. All FDISOs shall complete an 8-hour initial training session.

2. FDISOs shall attend eight of twelve proficiency training sessions offered in conjunction with Safety Committee meetings.

3. Operating as a safety officer at department live-fire training sessions, disaster drills, and full-scale HazMat scenarios shall be allowed to fill half of the annual proficiency training requirement.

4. FDISOs shall pursue additional training in topic-specific areas such as building construction, rehab and firefighter performance, tactics and strategies, fire behavior, and accident investigation.

Duties and Responsibilities

1. The FDISO shall maintain all assigned equipment and ensure that he or she has the appropriate forms, checklists, and other tools prior to responding to incidents as the FDISO.

2. The FDISOs are responsible to establish and maintain a duty system that ensures that a safety officer will be available for response to incidents.

3. When responding to an incident, the duty ISO shall use his or her normal radio designation (C/O, F/C, Chief, etc.) Upon arrival at an incident, the duty ISO shall make face-to-face contact with the IC. The IC shall confirm the ISO assignment.

4. The FDISO shall be responsible for incident duties outlined in standard operating guideline (SOG) #_____, Use of Safety Officer at Incidents.

5. The FDISO shall ensure that applicable follow-up care, investigation, and documentation is completed following incident scene injuries.

6. The FDISO shall initiate applicable procedures for a firefighter line-of-duty death and be the lead officer in processing the Public Safety Officer's Benefit (PSOB) application.

7. The FDISO shall work with the IC and the Department Training Officer in developing an incident critique and appropriate recommendations for follow-up training.

Appendix

B

Sample Incident Safety Officer Standard Operating Guideline (Use of Safety Officer at Incidents)

PURPOSE

To provide guidance and procedure for the utilization of a safety officer at designated and significant incidents.

RESPONSIBILITY

It shall be the responsibility of Incident Commanders (ICs) to designate a Fire Department Incident Safety Officer (FDISO) at incidents suggested in this standard operating guideline (SOG). It shall be the responsibility of designated FDISOs to follow procedure contained here within. It shall be the responsibility of all on-scene fire personnel to work with the FDISO to recognize and minimize risks associated with incident environments and operations.

PROCEDURE

General Guidelines

1. The department recognizes that certain incidents present a significant or increased risk to firefighters. With these incidents come an increased responsibility to monitor firefighting actions and environmental conditions. The appointment of a Fire Department Incident Safety Officer can increase an Incident Commander's effectiveness in protecting firefighters.

2. The Incident Commander shall appoint an FDISO early during an incident in order to maximize the effectiveness of the IC/FDISO team. The FDISO shall don a high-visibility "Safety Officer" vest as a means to signify to all personnel the presence of the Safety Officer. Supervisors and leaders shall report hazards to the FDISO in the course of operations.

3. An appointed Fire Department Incident Safety Officer shall have the authority of the Incident Commander to stop or alter any operation, action, or personal exposure that presents a threat to the life safety of a firefighter, crew, or liaison person.

4. Any changed, altered, or stopped assignment made by the FDISO shall be immediately communicated to the Incident Commander.

5. The FDISO shall have the authority to appoint assistant safety officers if the size, scope, and duration of the incident warrants additional assistance.

6. The FDISO shall possess a two-way radio and monitor radio transmissions. Imminent hazards shall be communicated to the Incident Commander and affected crews upon discovery.

Automatic FDISO Response

The following types of dispatched incidents are cause for the FDISO to respond to the incident:

1. Aircraft Alert 2 and 3 involving index B or larger aircraft. Also, any aircraft incident outside the airport critical zones.

2. Reported fire in a commercial structure. Included are reports of heavy smoke or fire from a commercial structure. Odor investigations, automatic fire alarms, and other investigative-type alarms are not included.

3. Activation of a specialty team for an incident. Included are rope rescue, heavy rescue, dive rescue, HazMat, and wildland team activations.

4. Incidents at "target" hazard complexes. This includes person trapped, collapses, smoke or odor reports, spills, and any other incident other than single patient medicals. Complexes include:

Merix Corporation	Plaza Apartments
Walmart Distribution Center	Woodward-Governor
Omni-Trax	Collins Plating
Good Samaritan	The Wexford
Colorado Crystal	McKee Medical Center

5. Any reported fire or working incident when climatic conditions become extreme. This includes:
   - Temperatures over 97°F
   - Temperatures below 0°F
   - Winds gusting over 60 mph
   - Snow depths over 12 inches
   - Wind-driven snow over 30 mph

6. The FDISO shall monitor incidents and self-dispatch if the communicated information suggests a difficult or significant incident (examples: A Motor Vehicle Accident (MVA) with a school bus involved, explosion, hostage situation, etc.)

Automatic FDISO Delegation

The Incident Commander shall delegate an officer to fill the FDISO position if any of the following conditions exist:

1. A second alarm is struck (or greater).
2. A firefighter injury requiring transport or a line-of-duty death.
3. Five or more group/division assignments or division of operations into branches.
4. Any incident where firefighters must take extreme risks or the Incident Commander wishes to delegate the safety responsibility.

Specific ISO Duties

1. The FDISO shall report to the IC and confirm the need for a safety officer prior to acting as the safety officer.
2. The FDISO shall utilize the FDISO checklist and prioritize safety efforts. Hazards found shall be communicated to the appropriate crew and the IC.
3. The FDISO shall take the lead in the investigation and documentation of firefighter injuries on scene.
4. When a significant safety violation is found, the FDISO shall take action or cause to take action to ensure that the violation is addressed and follow-up is documented at the company level.

5. Upon hearing a "MAYDAY," or witnessing a collapse, or noting any other firefighter emergency condition, the FDISO shall immediately report to the IC and assist with developing a systematic approach to the rescue and recovery of firefighters.

6. If the FDISO is to perform any reconnaissance or operation within the hot zone, the FDISO shall team up with a partner and be tracked through the accountability system.

7. The FDISO shall refrain from assisting with any task assignment and should maintain an observer role in order to best provide consultant information to the Incident Commander. If task involvement is required to prevent a firefighter injury, this action may take place, however, the task should be altered once the immediate threat is abated.

Appendix

C

Incident Safety Officer Checklist

UPON ARRIVAL or ASSIGNMENT
- ☐ Don Safety Officer vest
- ☐ Obtain briefing from Incident Commander including:
  - SITSTAT/RESTAT
  - Action Plan
  - Known hazards or concerns
- ☐ Prioritize Safety Officer duties:
  - Risk review
  - Reconnaissance
  - Resource evaluation
  - Reporting/planning
  - (each of these are listed below)
- ☐ Start tracking incident elapsed time
- ☐ Determine need and request additional safety officers

RISK REVIEW
- ☐ Define risk level for Action Plan
  - life at risk
  - property at risk
  - mitigation only
- ☐ Determine frequency and severity of hazards
- ☐ Prioritize hazard control recommendations
- ☐ Address unacceptable risk situations
  - stop or alter if life threatening
  - immediately advise IC

RECONNAISSANCE—Environmental
- ☐ Perform 360° scene survey
- ☐ Define principle hazard and location
  - validate hot/warm/cold zones
- ☐ Evaluate integrity of environment
  - stable—not likely to change
  - stable—may change
  - starting to change
  - rapidly changing
- ☐ Classify structure(s) involved
  - construction type
  - materials used
  - loads imposed

Verbally Confirm Construction Type and Hazards w/IC
- ☐ Evaluate collapse potential
  - structural degradation
  - excessive loads
  - scope of collapse
  - stability profile after collapse
- ☐ Define scope of utility involvement

- ☐ Evaluate effects of weather on incident
- ☐ Identify access/egress routes and deficiencies
- ☐ Define traffic hazards

RECONNAISSANCE—Operational
- ☐ Observe tactical assignments
  - tactical effectiveness
  - team effectiveness
  - tool application/effectiveness
  - action plan compatibility
- ☐ Check exposure of teams
  - correct PPE
  - aware of hazard
  - appropriate risk level
  - escape routes
- ☐ Determine injury potential
  - fall hazards
  - rehab profile
- ☐ Evaluate apparatus placement/exposure
- ☐ Monitor radio communications
  - dysfunction
  - status reports
  - MAYDAY report

RESOURCES
- ☐ Check scene attendance
  - too few/too many
  - plans for additional
- ☐ Determine rapid intervention status/capability
- ☐ Determine number of crews at risk
- ☐ Check effectiveness of accountability system
- ☐ Start time-pacing incident
  - dispatch assistance
  - anticipate total on-scene time
- ☐ Evaluate rehab process and effectiveness
- ☐ Assess the need for CISM

REPORT/PLANNING
- ☐ Communicate concerns to IC
- ☐ Develop contingencies
- ☐ Attend planning meetings
- ☐ Review action plan revisions/updates
- ☐ Prepare hazard awareness and safety briefs for arriving crews

AFTER THE INCIDENT
- ☐ Prepare a report of still-existing hazards
- ☐ Document Safety Officer actions
- ☐ Provide information to the IC and Training Officer for incident critique
- ☐ Complete accident/injury documentation IAW EOs

Acronyms

ACLS	Advanced cardiac life support
AFFF	Aqueous film forming foam
ALS	Advanced life support
ANSI	American National Standards Institute
ARFF	Aircraft rescue and firefighting
AVL	Automatic vehicle locators
BLEVE	Boiling liquid expanding vapor explosion
BLS	Basic life support
CAD	Computer-aided dispatch
CFR	Code of Federal Regulations
CISD	Critical incident stress debriefing
CISM	Critical incident stress management
dBA	decibel level
DSO	Duty safety officer
EAP	Employee assistance programs
EMS	Emergency medical services
EO	Emergency operations
ESO	Emergency safety officer
FAST	Firefighter assistance and search team
FDISO	Fire Department Incident Safety Officer
FDNY	New York City Fire Department
FEMA	Federal Emergency Management Administration
FGC	Fireground commander
FIRESCOPE	Fire Resources of Southern California Organized for Potential Emergencies
GFI	Ground fault interruption
GO	General operations
GPS	Global positioning systems
HazMat	Hazardous Materials
HAZWOPER	Hazardous Waste Operations and Emergency Response
HSO	Health and Safety Officer
HVAC	Heating, ventilation, and air-conditioning
IAFF	International Association of Firefighters
IAW	In accordance with
IC	Incident Commander
ICS	Incident command system
IDLH	Immediately dangerous to life and health
IFSTA	International Fire Service Training Association
ISO	Incident Safety Officer (now FDISO)
ISSO	Incident Scene Safety Officer
I-Zone	Interface zone
LZ	Landing zone
MSDS	Material safety data sheet
MVA	Motor Vehicle Accident
NCCI	National Council on Compensation Insurance
NFPA	National Fire Protection Association
NIIMS	National Interagency Incident Management System
NIOSH	National Institute for Occupational Safety and Health
NOAA	National Oceanic and Aeronautic Association
OSHA	Occupational Safety and Health Administration

PAR	Personnel accountability report	SO	Safety officer
PAS	Personal alert system	SOF1	Safety Officer Type 1
PIA	Postincident analysis	SOF2	Safety Officer Type 2
PPE	Personal protective equipment	SOG	Standard operating guidelines
PSOB	Public safety officer's benefit	SOP	Standard operating procedure
RAT	Rapid action team	TSO	Training/Safety Officer
RECON	Reconnaissance	UL	Underwriters' Laboratories
RIC	Rapid intervention crews	USAR	Urban search and rescue
SCBA	Self-contained breathing apparatus		

Glossary

Accident chain A series of events and conditions that can lead or have led to an accident. These events and conditions are typically classified into five areas: environment, human factors, equipment, event, and inventory.

Accountability system A system that readily identifies the location and function/assignment of all personnel operating at an incident.

Action plan The objectives that help achieve the desired strategy, tactics, risk management, and safety outcomes for an incident. The action plan is approved by the Incident Commander and updated as needed.

Defensible space A fire service term referring to the clear fuel space around a structure. Defensible space is, at minimum, 30 feet of cleared area that allows for the defense of a structure during wildland and I-zone fires.

Ergonomics The science that seeks to adapt work conditions to the worker in order to prevent injuries.

Fire flow The amount of water, expressed in gallons per minute (gpm), that needs to be applied to a fire in order to absorb the heat released from burning fuels. If the gpm rate meets fire flow for a given fire, the fire will be knocked down, allowing extinguishment.

FDISO Fire Department Incident Safety Officer. Title given to the officer assigned by the incident commander to handle safety responsibilities and duties.

FDISO Action Model A four-area thought processing model that assists the FDISO in remembering items that must be evaluated during emergency incidents.

HazMat Fire service short word for hazardous materials.

Health and Safety Officer (HSO) Title given to the person or persons who manage and administer the occupational safety and health program for a fire department.

Hot Zone The area immediately surrounding and including an environment that is immediately dangerous to life and health.

Imminent danger A situation that is immediately dangerous to life and health. Usually warrants action to avoid an injury. This action can take the form of terminating, altering, or withdrawing from the task or situation.

Incident safety officer (ISO) Title given to the officer assigned by the incident commander to handle safety responsibilities and duties. The person may or may not be the fire department's health and safety officer. Because the acronym ISO has been traditionally used by the Insurance Services Office, this text refers to the ISO as the FDISO.

I-zone Term given to the wildland/urban interface. The geographic areas where homes and businesses blend in with hill or mountain areas, which present significant fire control problems for fire departments. These problems include restricted or steep access, dense fuels that may be subject to dryness, limited water supply, and wind-pushed fire spread.

loom-up A large quantity of smoke that can be seen rising from a distance, usually indicative of a working, hostile fire.

Mastery The concept that an individual can achieve 90% of an objective 90% of time.

Microburst A sudden downburst of winds and precipitation that can produce straight-line winds of over 100 mph.

Mitigation   Actions taken to eliminate a hazard or make the hazard less severe or less likely to cause harm.

National Council on Compensation Insurance (NCCI)   A group responsible for setting insurance rates, specifically worker's compensation insurance, for various work groups or classifications.

National Interagency Incident Management System (NIIMS)   A federally-recognized system for managing multijurisdictional emergency incidents. The National Wildfire Coordinating Group uses NIIMS as its incident management system. NIIMS outlines specific titles and duties for each position of the system.

PPE   Acronym for personal protective equipment

Polyphasic   A high-tech term for multitasking or multipriority thinking.

PAR   Acronym for personnel accountability report. This is a crew-by-crew radio report to determine if all firefighters are accounted for. A PAR may be called for at given intervals during an incident or after a fireground event such as a building collapse.

Postincident analysis (PIA)   A formal or informal critique session following an incident. Typically, operational successes, outcomes, and problem areas are brought to light to avoid future problems and to reinforce the use of accepted policy and procedure.

Potential danger   A situation that, given time, proximity, or subtle change, can cause injury. Persons operating near a potential danger should exercise caution in the form of observation, distancing, or constant evaluation.

Pyrolysis   The chemical breakdown of a material through heating.

Recon   A shortened form of the word *reconnaissance*. Reconnaissance is the survey or examination that seeks out critical information.

Risk management   The process of minimizing the chance, degree, or probability of damage, loss, or injury.

Safety officer   A generic title that can represent either an incident safety officer or a health and safety program officer or manager. Specificity is preferred and the terms health and safety officer (HSO) and incident safety officer (ISO) have been chosen by the National Fire Academy and NFPA.

Topography   The accurate description of a place or the lay of the land including hills, reliefs, rivers, lakes, and man-made land features such as dams and bridges.

USAR teams   Urban search and rescue teams. These are highly trained and specially equipped teams, recognized by the Federal Emergency Management Administration (FEMA), to respond to significant structural collapses resulting from earthquakes, terrorist activities, and other events. Most USAR teams are prepared to deploy to other jurisdictions via truck or aircraft.

Index

Vehicle integrity, 139–142
  access to vehicle interior, 142
  stability and position, 139–141
  status of vehicle systems, 141–142
    electrical system, 142
    fuel system, 141–142
Visibility, maintaining, 82

Water system integrity, 138–139
Weak link, in buildings, 70
  connections, 70
  overloading, 71
  trusses, 70, 71
Weather, 62
  extremes, 46
  humidity, 145–146
  potential for change or storms, 147–148

  temperature, 146–147
  and wildland fires, 131
  wind, 145, 146
Wildland fires, 62, 130
  fuels involved, 130–131
    continuity of fuels, 130
    moisture content, 131
    type of fuel, 130
  location of fire, 130
  weather, 131
Wildland interface fire, 43–44
Wind, 145, 146
Work areas, 31
Worker's compensation, 11–12
Worker trap, 87–88
Working incident, 47